Published by Coordination Group Publications Ltd.

From original material by Richard Parsons.

Editors:
Ellen Bowness, Tom Cain, Katherine Craig, Gemma Hallam, Sarah Hilton, Sharon Keeley,
Andy Park, Rose Parkin, Kate Redmond, Katherine Reed, Alan Rix, Rachel Selway,
Ami Snelling, Claire Thompson, Julie Wakeling.

Contributors:
James Foster, Sandy Gardner, Julian Hardwick, Derek Harvey, Steven Phillips,
Claire Reed, Adrian Schmit, Claire Stebbing.

ISBN: 978 1 84146 638 5

Groovy website: www.cgpbooks.co.uk

Printed by Elanders Ltd, Newcastle upon Tyne.
Jolly bits of clipart from CorelDRAW®

Theories Come, Theories Go

SCIENTISTS ARE ALWAYS RIGHT — OR ARE THEY?

Well, it'd be nice if that were so, but it just isn't — never has been and never will be.
Increasing scientific knowledge involves making mistakes along the way. Let me explain...

Scientists Come Up with Hypotheses — Then Test Them

1) Scientists try and explain things. Everything.

2) They start by observing or thinking about something they don't understand — it could be anything, e.g. planets in the sky, a person suffering from an illness, what matter is made of... anything.

Hundreds of years ago, we thought demons caused illness.

3) Then, using what they already know (plus a bit of insight), they come up with a hypothesis (a theory) that could explain what they've observed. But a hypothesis is just a theory, a belief. And believing something is true doesn't make it true — not even if you're a scientist.

4) So the next step is to try and convince other scientists that the hypothesis is right — which involves using evidence. First, the hypothesis has to fit the evidence already available — if it doesn't, it'll convince no one.

5) Next, the scientist might use the hypothesis to make a prediction — a crucial step. If the hypothesis predicts something, and then evidence from experiments backs that up, that's pretty convincing. This doesn't mean the hypothesis is true (the 2nd prediction, or the 3rd, 4th or 25th one, might turn out to be wrong) — but a hypothesis that correctly predicts something in the future deserves respect.

Other Scientists Will Test the Hypotheses Too

Then we thought it was caused by 'bad blood' (and treated it with leeches).

1) Now then... other scientists will want to use the hypothesis to make their own predictions, and they'll carry out their own experiments. (They'll also try to reproduce earlier results.) And if all the experiments in all the world back up the hypothesis, then scientists start to have a lot of faith in it.

2) However, if a scientist somewhere in the world does an experiment that doesn't fit with the hypothesis (and other scientists can reproduce these results), then the hypothesis is in trouble. When this happens, scientists have to come up with a new hypothesis (maybe a modification of the old theory, or maybe a completely new one).

3) This process of testing a hypothesis to destruction is a vital part of the scientific process. Without the 'healthy scepticism' of scientists everywhere, we'd still believe the first theories that people came up with — like thunder being the belchings of an angered god (or whatever).

If Evidence Supports a Hypothesis, It's Accepted — for Now

1) If pretty much every scientist in the world believes a hypothesis to be true because experiments back it up, then it usually goes in the textbooks for students to learn.

Now we know most illnesses are due to microorganisms.

2) Our currently accepted theories are the ones that have survived this 'trial by evidence' — they've been tested many, many times over the years and survived (while the less good ones have been ditched).

3) However... they never, never become hard and fast, totally indisputable fact. You can never know... it'd only take one odd, totally inexplicable result, and the hypothesising and testing would start all over again.

You expect me to believe that — then show me the evidence...

If scientists think something is true, they need to produce evidence to convince others — it's all part of testing a hypothesis. One hypothesis might survive these tests, while others won't — it's how things progress. And along the way some hypotheses will be disproved — i.e. shown not to be true. So, you see... not everything scientists say is true. It's how science works.

Your Data's Got to Be Good

Evidence is the key to science — but not all evidence is equally good.
The way evidence is gathered can have a big effect on how trustworthy it is...

Lab Experiments Are Better Than Rumour or Small Samples

1) Results from controlled experiments in laboratories are great. A lab is the easiest place to control variables so that they're all kept constant (except for the one you're investigating). This makes it easier to carry out a fair test. It's also the easiest way for different scientists around the world to carry out the same experiments. (There are things you can't study in a lab though, like climate.)

2) Old wives' tales, rumours, hearsay, "what someone said", and so on, should be taken with a pinch of salt. They'd need to be tested in controlled conditions to be genuinely scientific.

3) Data based on samples that are too small don't have much more credibility than rumours do. A sample should be representative of the whole population (i.e. it should share as many of the various characteristics in the population as possible) — a small sample just can't do that.

Evidence Is Only Reliable If Other People Can Repeat It

Scientific evidence needs to be reliable (or reproducible). If it isn't, then it doesn't really help.

RELIABLE means that the data can be reproduced by others.

In 1989, two scientists claimed that they'd produced 'cold fusion' (the energy source of the Sun — but without the enormous temperatures). It was huge news — if true, this could have meant energy from sea water — the ideal energy solution for the world... forever. However, other scientists just couldn't get the same results — i.e. the results weren't reliable. And until they are, 'cold fusion' isn't going to be generally accepted as fact.

Evidence Also Needs to Be Valid

To answer scientific questions scientists often try to link changes in one variable with changes in another. This is useful evidence, as long as it's valid.

VALID means that the data is reliable AND answers the original question.

EXAMPLE: DO POWER LINES CAUSE CANCER?
Some studies have found that children who live near overhead power lines are more likely to develop cancer. What they'd actually found was a correlation between the variables "presence of power lines" and "incidence of cancer" — they found that as one changed, so did the other. But this evidence is not enough to say that the power lines cause cancer, as other explanations might be possible. For example, power lines are often near busy roads, so the areas tested could contain different levels of pollution from traffic. Also, you need to look at types of neighbourhoods and lifestyles of people living in the tested areas (could diet be a factor... or something else you hadn't thought of...).
So these studies don't show a definite link and so don't answer the original question.

Controlling All the Variables Is Really Hard

In reality, it's very hard to control all the variables that might (just might) be having an effect. You can do things to help — e.g. choose two groups of people (those near power lines and those far away) who are as similar as possible (same mix of ages, same mix of diets etc). But you can't easily rule out every possibility. If you could do a properly controlled lab experiment, that'd be better — but you just can't do it without cloning people and exposing them to things that might cause cancer... hardly ethical.

Does the data really say that?...

If it's so hard to be definite about anything, how does anybody ever get convinced about anything? Well, what usually happens is that you get a load of evidence that all points the same way. If one study can't rule out a particular possibility, then maybe another one can. So you gradually build up a whole body of evidence, and it's this (rather than any single study) that convinces people.

Bias and How to Spot It

Scientific results are often used to help people make a point (e.g. politicians, environmental campaigners... and so on). But results are sometimes presented in a biased way — and you need to be able to spot that.

You Don't Need to Lie to Make Things Biased

1) For something to be misleading, it doesn't have to be untrue. We tend to read scientific facts and assume that they're the 'truth', but there are many different sides to the truth. Look at this headline...

> **Scientists say 1 in 2 people are of above average weight**

Sounds like we're a nation of fatties. It's a scientific analysis of the facts, and almost certainly true.

2) But an average is a kind of 'middle value' of all your data.
Some readings are higher than average (about half of them, usually).
Others will be lower than average (the other half).

So the above headline (which made it sound like we should all lose weight) could just as accurately say:

> **Scientists say 1 in 2 people are of below average weight**

3) The point is... both headlines sound quite worrying, even though they're not. That's the thing... you can easily make something sound really good or really bad — even if it isn't. You can...

① ...use only some of the data, rather than all of it:

"Many people lost weight using the new SlimAway diet. Buy it now!!"

"Many" could mean anything — e.g. 50 out of 5000 (i.e. 1%). But that could be ignoring most of the data.

② ...phrase things in a 'leading' way: 90% fat free!

Would you buy it if it were "90% cyanide free"? That 10% is the important bit, probably.

③ ...use a statistic that supports your point of view:

| The amount of energy wasted is increasing. | Energy wasted per person is decreasing. | The rate at which energy waste is increasing is slowing down. |

These describe the same data. But two sound positive and one negative.

Think About Why Things Might Be Biased

1) People who want to make a point can sometimes present data in a biased way to suit their own purposes (sometimes without knowing they're doing it).

2) And there are all sorts of reasons why people might want to do this — for example...

- Governments might want to persuade voters, other governments, journalists, etc. Evidence might be ignored if it could create political problems, or emphasised if it helps their cause.
- Companies might want to 'big up' their products. Or make impressive safety claims, maybe.
- Environmental campaigners might want to persuade people to behave differently.

3) People do it all the time. This is why any scientific evidence has to be looked at carefully. Are there any reasons for thinking the evidence is biased in some way?

- Does the experimenter (or the person writing about it) stand to gain (or lose) anything? (For example, are they being funded by a particular company or group?)
- Might someone have ignored some of the data for political or commercial reasons?
- Is someone using their reputation rather than evidence to help make their case?

Tell me what you want people to believe, and I'll find a statistic to help...

So scientific data and the person presenting it need to be looked at carefully. That doesn't mean the scientific data's always misleading, just that you need to be careful. The most credible argument will be the one that describes all the data that was found, and gives the most balanced view of it.

Science Has Limits

Science can give us amazing things — cures for diseases, space travel, heated toilet seats...
But science has its limitations — there are questions that it just can't answer.

Some Questions Are Unanswered by Science — So Far

1) We don't understand everything. And we never will. We'll find out more, for sure — as more hypotheses are suggested, and more experiments are done. But there'll always be stuff we don't know.

   For example, today we don't know as much as we'd like about climate change (global warming). Is climate change definitely happening? And to what extent is it caused by humans?

2) These are complicated questions and at the moment scientists don't all agree on the answers. But eventually, we probably will be able to answer these questions once and for all.

3) By then, though, there'll be loads of new questions to answer.

Other Questions Are Unanswerable by Science

1) Then there's the other type... questions that all the experiments in the world won't help us answer — the "Should we be doing this at all?" type questions. There are always two sides...

2) Take embryo screening (which allows you to choose an embryo with particular characteristics). It's possible to do it — but does that mean we should?

3) Different people have different opinions. For example...

- Some people say it's good... couples whose existing child needs a bone marrow transplant, but who can't find a donor, will be able to have another child selected for its matching bone marrow. This would save the life of their first child — and if they want another child anyway... where's the harm?

- Other people say it's bad... they say it could have serious effects on the child. In the above example the new child might feel unwanted — thinking they were only brought into the world to help someone else. And would they have the right to refuse to donate their bone marrow (as anyone else would)?

4) This question of whether something is morally or ethically right or wrong can't be answered by more experiments — there is no 'right' or 'wrong' answer.

5) The best we can do is get a consensus from society — a judgement that most people are more or less happy to live by. Science can provide more information to help people make this judgement, and the judgement might change over time. But in the end it's up to people and their consciences.

Loads of Other Factors Can Influence Decisions Too

Here are some other factors that can influence decisions about science, and the way science is used:

Economic factors:
- Companies very often won't pay for research unless there's likely to be a profit in it.
- Society can't always afford to do things scientists recommend without cutting back elsewhere (e.g. investing heavily in alternative energy sources).

Social factors:
- Decisions based on scientific evidence affect people — e.g. should fossil fuels be taxed more highly (to invest in alternative energy)? Should alcohol be banned (to prevent health problems)? Would the effect on people's lifestyles be acceptable...

Environmental factors:
- Genetically modified crops may help us produce more food — but some people say they could cause environmental problems (see page 43).

Science is a 'real-world' subject...

Science isn't just done by people in white coats in labs who have no effect on the outside world. Science has a massive effect on the real world every day, and so real-life things like money, morals and how people might react need to be considered. It's why a lot of issues are so difficult to solve.

The Nervous System

First thing on the menu is a page about the nervous system. The nervous system is what lets you react to what goes on around you, so you'd find life tough without it.

Sense Organs Detect Stimuli

A stimulus is a change in your environment which you may need to react to (e.g. a recently pounced tiger). You need to be constantly monitoring what's going on so you can respond if you need to.

1) You have five different sense organs — eyes, ears, nose, tongue and skin.

2) They all contain different receptors. Receptors are groups of cells which are sensitive to a stimulus. They change stimulus energy (e.g. light energy) into electrical impulses.

3) A stimulus can be light, sound, touch, pressure, chemical, or a change in position or temperature.

> Sense organs and Receptors
> Don't get them mixed up:
>
> The eye is a sense organ — it contains light receptors.
> The ear is a sense organ — it contains sound receptors.

The Five Sense Organs and the receptors that each contains:

1) Eyes — Light receptors.

2) Ears — Sound and "balance" receptors.

3) Nose — Smell receptors — sensitive to chemical stimuli.

4) Tongue — Taste receptors: — sensitive to bitter, salt, sweet and sour, plus the taste of savoury things like monosodium glutamate (MSG) — chemical stimuli.

5) Skin — Sensitive to touch, pressure and temperature change.

Sensory Neurones

The nerve cells that carry signals as electrical impulses from the receptors in the sense organs to the central nervous system.

Motor Neurones

The nerve cells that carry signals to the effector muscles or glands.

Effectors

Muscles and glands are known as effectors. They respond in different ways — muscles contract in response to a nervous impulse, whereas glands secrete hormones.

The Central Nervous System Coordinates a Response

1) The central nervous system (CNS) is where all the information from the sense organs is sent, and where reflexes and actions are coordinated.

The central nervous system consists of the brain and spinal cord only.

2) Neurones (nerve cells) transmit the information (as electrical impulses) very quickly to and from the CNS.

3) "Instructions" from the CNS are sent to the effectors (muscles and glands), which respond accordingly.

Your tongue's evolved for Chinese meals — sweet, sour, MSG...

Listen up... the thing with GCSE Science is that it's not just a test of what you know — it's also a test of how well you can apply what you know. For instance, you might have to take what you know about a human and apply it to a horse (easy... sound receptors in its ears, light receptors in its eyes, etc.), or to a snake (so if you're told that certain types of snakes have heat receptors in nostril-like pits on their head, you should be able to work out what type of stimulus those pits are sensitive to). Thinking in an exam... gosh.

Reflexes

Your brain can <u>decide</u> how to respond to a stimulus <u>pretty quickly</u>.
But sometimes waiting for your brain to make a decision is just <u>too slow</u>. That's why you have <u>reflexes</u>.

Reflexes **Help** Prevent Injury

1) <u>Reflexes</u> are <u>automatic</u> responses to certain stimuli — they can reduce the chances of being injured.

2) For example, if someone shines a <u>bright light</u> in your eyes, your <u>pupils</u> automatically get smaller so that less light gets into the eye — this stops it getting <u>damaged</u>.

3) Or if you get a shock, your body releases the <u>hormone</u> adrenaline automatically — it doesn't wait for you to <u>decide</u> that you're shocked.

4) The route taken by the information in a reflex (from receptor to effector) is called a <u>reflex arc</u>.

The Reflex Arc Goes Through **the** Central Nervous System

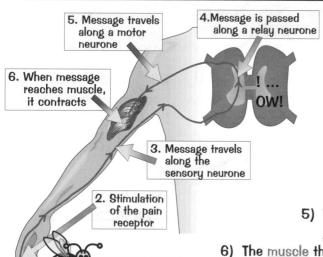

5. Message travels along a motor neurone

4. Message is passed along a relay neurone

6. When message reaches muscle, it contracts

! ... OW!

3. Message travels along the sensory neurone

2. Stimulation of the pain receptor

1. Cheeky bee stings finger

1) The neurones in reflex arcs go through the <u>spinal cord</u> or through an <u>unconscious part of the brain</u>.

2) When a <u>stimulus</u> (e.g. a painful bee sting) is detected by receptors, an impulse is sent along a <u>sensory neurone</u> to the CNS.

3) In the CNS the sensory neurone passes on the message to another type of neurone — a <u>relay neurone</u>.

4) Relay neurones <u>relay</u> the impulse to a <u>motor neurone</u>.

5) The impulse then travels along the motor neurone to the <u>effector</u> (in this example it's a muscle).

6) The <u>muscle</u> then <u>contracts</u> and moves your hand away from the bee.

7) Because you don't have to think about the response (which takes time) it's <u>quicker</u> than normal responses.

Here's a <u>block diagram</u> of a <u>reflex arc</u> — it shows what happens, from stimulus to response.

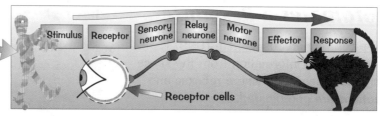

Stimulus | Receptor | Sensory neurone | Relay neurone | Motor neurone | Effector | Response

Receptor cells

Synapses **Connect** Neurones

1) <u>Neurones</u> transmit information as <u>electrical impulses</u> around the body. Neurones have <u>branched endings</u> so they can <u>connect</u> with lots of other neurones. And they're <u>long</u>, which <u>speeds up</u> the impulse (fewer connections means a quicker signal).

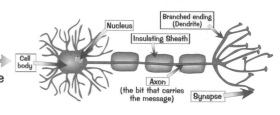

Nucleus

Branched ending (Dendrite)

Insulating Sheath

Cell body

Axon (the bit that carries the message)

Synapse

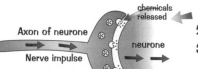

chemicals released

Axon of neurone

neurone

Nerve impulse

2) The <u>connection</u> between <u>two neurones</u> is called a <u>synapse</u>.

3) The nerve signal is transferred by <u>chemicals</u> which <u>diffuse</u> (move) across the gap.

4) These chemicals set off a <u>new electrical signal</u> in the <u>next</u> neurone.

Don't get all twitchy — just learn it...

The difference between a <u>reflex</u> and a "<u>considered response</u>" is the involvement of the conscious part of your brain. Reflexes may bypass your brain completely — your body just gets on with things.

The Eye

The <u>eye</u> is a good example of a <u>sense organ</u>, and you can see plenty <u>reflexes</u> in action there too...

Learn the Eye with All Its Labels:

1) The <u>cornea refracts</u> (bends) light into the eye. The <u>iris</u> controls <u>how much light</u> enters the <u>pupil</u> (<u>hole</u> in the <u>middle</u>). And the <u>lens</u> <u>focuses</u> the <u>light</u> onto the <u>retina</u> (the <u>light-sensitive</u> part — it's covered in light receptors called <u>rods</u> and <u>cones</u>).

2) <u>Rods</u> are more sensitive in <u>dim light</u> but <u>can't</u> sense colour. <u>Cones</u> are sensitive to <u>colours</u> but aren't so good in dim light (<u>red-green colour blindness</u> is due to a <u>lack</u> of certain <u>cone cells</u>).

3) The <u>optic nerve</u> carries impulses from the receptors to the <u>brain</u>.

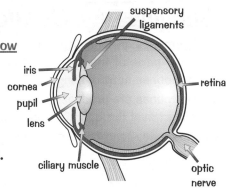

The Iris Reflex — Adjusting for Bright Light

<u>Very bright</u> light can <u>damage</u> the retina — so you have a reflex to protect it.

1) Very bright light triggers a reflex that makes the pupil <u>smaller</u>, allowing less light in. (See p6 for more about reflexes... but basically, in this case, <u>light receptors</u> detect the bright light, send a message to an unconscious part of the brain along a <u>sensory neurone</u>, and then the brain sends a message straight back along a <u>motor neurone</u> telling the <u>circular muscles</u> in the iris to <u>contract</u>, which makes the pupil smaller.)

2) The opposite process happens in dim light. This time, the brain tells the <u>radial muscles</u> to contract, which makes the pupil bigger.

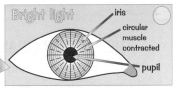

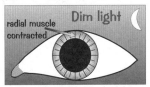

Focusing on Near and Distant Objects — Another Reflex

The eye focuses light by <u>changing</u> the <u>shape</u> of the <u>lens</u> — this is known as <u>accommodation</u>.

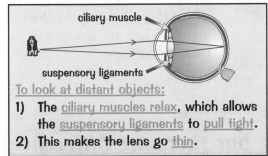

To look at distant objects:
1) The <u>ciliary muscles relax</u>, which allows the <u>suspensory ligaments</u> to <u>pull tight</u>.
2) This makes the lens go <u>thin</u>.

To look at near objects:
1) The <u>ciliary muscles contract</u>, which <u>slackens</u> the <u>suspensory ligaments</u>.
2) The lens becomes <u>fat</u>.

As you get older, your eye's <u>lens</u> loses <u>flexibility</u>, so it can't easily spring back to a round shape. This means light can't be <u>focused</u> well for near viewing, so older people often have to use reading glasses.

1) <u>Long-sighted</u> people are <u>unable to focus</u> on <u>near</u> objects. This occurs when the <u>cornea</u> or <u>lens</u> doesn't <u>bend</u> the light enough or the <u>eyeball</u> is too <u>short</u>. The images of near objects are brought into focus <u>behind</u> the <u>retina</u>.

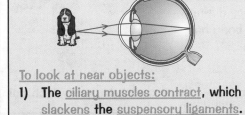

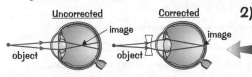

2) <u>Short-sighted</u> people are <u>unable to focus</u> on <u>distant</u> objects. This occurs when the <u>cornea</u> or <u>lens</u> bends the light <u>too much</u> or the <u>eyeball</u> is too <u>long</u>. The images of distant objects are brought into focus <u>in front</u> of the <u>retina</u>.

Binocular Vision Lets You Judge Depth

Some animals, including humans, have two eyes which <u>work together</u> — this is <u>binocular vision</u>. The brain uses small differences between what each eye sees to <u>judge distances</u> and <u>how fast</u> things are moving. It's handy for <u>catching prey</u> and deciding if it's safe to cross a road.

I think I'm a little long-sighted...

To see how important <u>binocular vision</u> is, cover one eye and try pouring water into a glass at arm's length. That's why you never see turkeys or horses pouring themselves a drink — no binocular vision.

Hormones

The other way to send information around the body (apart from along nerves) is by using hormones.

Hormones Are Chemical Messengers Sent in the Blood

1) Hormones are chemicals released directly into the blood. They're carried in the blood plasma to other parts of the body, but only affect particular cells (called target cells) in particular places. Hormones control things in organs and cells that need constant adjustment.

2) Hormones are produced in various glands, as shown on the diagram. They travel through your body at "the speed of blood".

3) Hormones tend to have relatively long-lasting effects.

Learn this definition:
HORMONES...
are chemical messengers
which travel in the blood
to activate target cells.

THE PITUITARY GLAND
This produces many important hormones including LH, FSH (see next page) and ADH (which controls water content).

PANCREAS
Produces insulin for the control of blood sugar (see p13).

OVARIES — females only
Produce oestrogen, which controls the menstrual cycle (see next page) and promotes all female secondary sexual characteristics during puberty, e.g. extra body hair and widening of hips.

Kidney

TESTES — males only
Produce testosterone, which promotes all male secondary sexual characteristics at puberty, e.g. extra hair and changes in body proportions.

These are just examples — there are loads more, each doing its own thing.

Hormones and Nerves Do Similar Jobs, but There Are Differences

NERVES: 1) Very FAST message.
2) Act for a very SHORT TIME.
3) Act on a very PRECISE AREA.

HORMONES: 1) SLOWER message.
2) Act for a LONG TIME.
3) Act in a more GENERAL way.

So if you're not sure whether a response is nervous or hormonal, have a think...

1) If the response is really quick, it's probably nervous. Some information needs to be passed to effectors really quickly (e.g. pain signals, or information from your eyes telling you about the lion heading your way), so it's no good using hormones to carry the message — they're too slow.

2) But if a response lasts for a long time, it's probably hormonal. For example, when you get a shock, a hormone called adrenaline is released into the bloodstream (causing the fight-or-flight response, where your body is hyped up ready for action). You can tell it's a hormonal response (even though it kicks in pretty quickly) because you feel a bit wobbly for a while afterwards.

Nerves, hormones — no wonder revision makes me tense...

Hormones control various organs and cells in the body, though they tend to control things that aren't immediately life-threatening. For example, they take care of most things to do with sexual development, pregnancy, birth, breast-feeding, blood sugar levels, water content... and so on. Pretty amazing really.

Puberty and the Menstrual Cycle

Hormones control almost everything to do with <u>sex</u> and <u>reproduction</u>.

Hormones *Promote* Sexual Characteristics at *Puberty*

At puberty your body starts releasing <u>sex hormones</u> — <u>testosterone</u> in men and <u>oestrogen</u> in women. These trigger off the <u>secondary sexual characteristics</u>:

In men —
1) <u>Extra hair</u> on face and body.
2) <u>Muscles develop</u>.
3) <u>Penis and testicles</u> enlarge.
4) <u>Sperm</u> production.
5) <u>Deepening</u> of <u>voice</u>.

In women —
1) <u>Extra hair</u> on underarms and pubic area.
2) <u>Hips widen</u>.
3) Development of <u>breasts</u>.
4) <u>Egg</u> release and <u>periods start</u>.

The *Menstrual Cycle* Has *Four Stages*

<u>Stage 1</u> <u>Day 1 is when bleeding starts</u>. The uterus lining breaks down for about four days.

<u>Stage 2</u> <u>The womb lining builds up again</u>, from day 4 to day 14, into a thick spongy layer full of blood vessels, ready to receive a fertilised egg.

<u>Stage 3</u> <u>An egg develops and is released</u> from the ovary at day 14.

<u>Stage 4</u> <u>The wall is then maintained</u> for about 14 days until day 28. If no fertilised egg has landed on the uterus wall by day 28, the spongy lining starts to break down and the whole cycle starts again.

It's *Controlled* by *Four Hormones*

1. FSH (follicle-stimulating hormone)
1) Produced in the <u>pituitary gland</u>.
2) Causes an <u>egg to develop</u> in one of the ovaries.
3) Stimulates the <u>ovaries</u> to produce <u>oestrogen</u>.

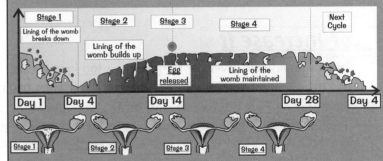

2. Oestrogen
1) Produced in the <u>ovaries</u>.
2) Causes the lining of the uterus to <u>thicken</u> and <u>grow</u>.
3) Stimulates the <u>production of LH</u> (which causes the release of an egg) and inhibits production of FSH.

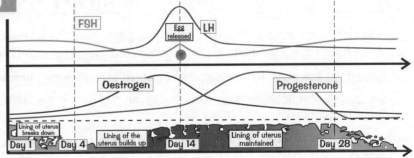

3. LH (luteinising hormone)
1) Produced by the <u>pituitary gland</u>.
2) Stimulates the <u>release of an egg</u> at day 14.

4. Progesterone
1) Produced in the <u>ovaries</u>.
2) <u>Maintains</u> the lining of the uterus. When the level of progesterone <u>falls</u>, the lining <u>breaks down</u>.

Which came first — the chicken or the luteinising hormone...

In the exam you might have to explain <u>what hormone causes what</u> in the menstrual cycle. It's tough, because they're all <u>interlinked</u>... but if you know your stuff, you should be okay. So learn it.

Controlling Fertility

The hormones FSH, oestrogen and LH can be used to change artificially how fertile a woman is.

Hormones Can Be Used to Reduce Fertility...

1) The hormone oestrogen can be used to prevent the release of an egg — so oestrogen can be used as a method of contraception. The pill is an oral contraceptive that contains oestrogen.

2) This may seem kind of strange (since naturally oestrogen helps stimulate the release of eggs). But if oestrogen is taken every day to keep the level of it permanently high, it inhibits the production of FSH, and after a while egg development and production stop and stay stopped.

Advantages
1) The pill's over 99% effective at preventing pregnancy.
2) It reduces the risk of getting some types of cancer.

Disadvantages
1) It isn't 100% effective — there's still a very slight chance of getting pregnant.
2) It can cause side effects like headaches, nausea, irregular menstrual bleeding, and fluid retention.
3) It doesn't protect against sexually transmitted infections (STIs).

...or Increase It

1) Some women have levels of FSH that are too low to cause their eggs to mature. This means that no eggs are released and the women can't get pregnant.

2) The hormone FSH can be taken by these women to stimulate egg production in their ovaries. (Well, in reality... FSH stimulates the ovaries to produce oestrogen, which stimulates the pituitary gland to produce LH, which stimulates the release of an egg... but FSH has the desired effect anyway.)

Advantage
It helps a lot of women to get pregnant when previously they couldn't... pretty obvious.

Disadvantages
1) It doesn't always work.
2) Too many eggs could be stimulated, resulting in unexpected multiple pregnancies (twins, triplets etc.).

IVF Can Also Help Couples to Have Children

IVF ("in vitro fertilisation") involves collecting eggs from the woman's ovaries and fertilising them in a lab using the man's sperm. These are then grown into embryos, which are transferred to the woman's uterus.

1) Hormones are given before egg collection to stimulate egg production (so several can be taken).

2) Oestrogen and progesterone are often given to make implantation of the embryo into the uterus more likely to succeed.

But the use of hormones in IVF can cause problems for some women...

1) Some women have a very strong reaction to the hormones — including abdominal pain, vomiting and dehydration.

2) There have been some reports of an increased risk of cancer due to the hormonal treatment (though other studies have reported no evidence of such a risk).

Too many initials to learn — FSH, IVF, STIs, CIA, DVD...

It's not just scientists who have an opinion on whether these hormonal treatments are good or bad... The teachings of several religions are interpreted by some people as being against contraception (though this often includes other methods too — not just hormone-based contraception). Some people also think that contraception increases promiscuous and irresponsible behaviour, since people know they are very unlikely to get pregnant (though they could still be at risk of being infected by a sexually transmitted infection (STI)). It's one of those situations where science can't provide all the answers.

Homeostasis

Homeostasis involves balancing body functions to maintain a "constant internal environment". Hormones are sometimes (but not always) involved.

Homeostasis is Maintaining a Constant Internal Environment

Conditions in your body need to be kept steady so that cells can function properly. This involves balancing inputs (stuff going into your body) with outputs (stuff leaving). For example...

1) Levels of CO_2 — respiration in cells (p60) constantly produces CO_2, which you need to get rid of.

2) Levels of oxygen — you need to replace the oxygen that your cells use up in respiration.

3) Water content — you need to keep a balance between the water you gain (in drink and food, and from respiration) and the water you pee, sweat and breathe out.

4) Body temperature — you need to get rid of excess body heat when you're hot, but retain heat when the environment is cold.

Negative Feedback Helps Keep All These Things Steady

NEGATIVE FEEDBACK

Changes in the environment trigger a response that counteracts the changes — e.g. a rise in body temperature causes a response that lowers body temperature.

This means that the internal environment tends to stay around a norm, which is the level at which the cells work best.

This only works within certain limits — if the environment changes too much then it might not be possible to counteract it (see the bit about heatstroke at the bottom of page 12).

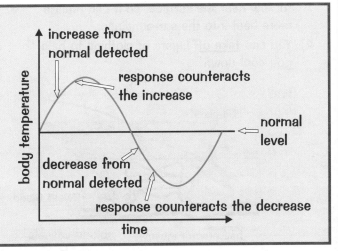

Water Is Lost from the Body in Various Ways

Water is taken into the body as food and drink and is lost from the body in the following ways:

1) through the **SKIN** as **SWEAT**...

2) via the **LUNGS** in **BREATH**...

3) via the kidneys as **URINE**.

Some water is also lost in faeces.

The balance between sweat and urine can depend on what you're doing, or what the weather's like...

On a **COLD DAY**, or when you're **NOT EXERCISING**, you don't sweat much, so you'll produce more urine, which will be pale (since the waste carried in the urine is more diluted).

On a **HOT DAY**, or when you're **EXERCISING**, you sweat a lot, and so you will produce less urine, but this will be more concentrated (and hence a deeper colour). You will also lose more water through your breath when you exercise because you breathe faster.

If you do enough revision, you can avoid negative feedback...

Negative feedback is a fancy-sounding name for a not-very-complicated idea. It's common sense really. For example, if you looked sad, I'd try and cheer you up. And if you looked really happy, I'd probably start to annoy you by flicking the backs of your ears. It stops things getting out of balance, I think.

Homeostasis

Homeostasis is a fancy word, but it covers lots of things, so maybe that's fair enough.

Body Temperature is Kept at About 37 °C

1) All <u>enzymes</u> work best at a certain temperature. The enzymes in the human body work best at about <u>37 °C</u> — and so this is the temperature your body tries to maintain.

2) A part of the <u>brain</u> acts as your own <u>personal thermostat</u>. It's sensitive to the blood temperature in the brain, and it <u>receives</u> messages from the skin that provide information about <u>skin temperature</u>.

3) To keep you at this temperature your body does these things:

When You're TOO HOT:

1) <u>Hairs</u> lie flat.

2) <u>Lots of sweat</u> is produced — when it <u>evaporates</u> it <u>transfers heat</u> from you to the environment, cooling you down.

3) <u>Blood vessels</u> close to the surface of the skin <u>widen</u> (vasodilation). This allows more blood to flow near the surface, so it can radiate more heat into the surroundings.

4) You can <u>take off</u> layers of <u>clothing</u> to help you cool down.

When You're TOO COLD:

1) <u>Hairs</u> stand on end to trap an insulating layer of air which helps keep you warm.

2) <u>Very little sweat</u> is produced.

3) <u>Blood vessels</u> near the surface <u>constrict</u> (vasoconstriction) so that less heat can be transferred from the blood to the surroundings.

4) You <u>shiver</u>, and the movement generates heat in the muscles. <u>Exercise</u> does the same.

5) You can put on <u>more clothes</u>, to trap the heat in.

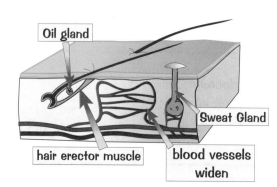

Oil gland — hair erector muscle — blood vessels widen — Sweat Gland

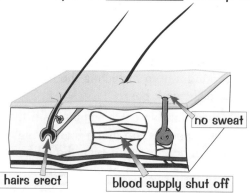

hairs erect — blood supply shut off — no sweat

If you're exposed to <u>high temperatures</u> you can get <u>dehydrated</u> and you could get <u>heat stroke</u>. This can <u>kill</u> you (see below).

Your body temperature can drop to dangerous levels if you're exposed to <u>very low temperatures</u> for a long time — this is called <u>hypothermia</u>. If you don't get help quickly you can <u>die</u>.

Ion Content Is Regulated by the Kidneys

1) <u>Ions</u> (e.g. sodium, Na$^+$) are taken into the body in <u>food</u>, then absorbed into the blood.

2) If the food contains <u>too much</u> of any kind of ion then the excess ions need to be <u>removed</u>. E.g. a salty meal will contain far too much Na$^+$.

3) Some ions are lost in <u>sweat</u> (which tastes salty, you'll have noticed).

4) The kidneys will <u>remove the excess</u> from the blood — this is then got rid of in <u>urine</u>.

Kidneys

5) <u>Sports drinks</u> (which usually contain <u>electrolytes</u> and <u>carbohydrates</u>) can help your body keep things in order. The <u>electrolytes</u> (e.g. sodium) replace those lost in <u>sweat</u>, while the carbohydrates can give a bit of an energy boost. But claims about sports drinks need to be looked at carefully.

Heat stroke is no laughing matter...

If you're in really high temperatures for a long time you can get <u>heat stroke</u> — <u>sweating stops</u>, since you get so <u>dehydrated</u>, and there's a <u>big rise</u> in your body temperature. If you don't cool down you can die. Fortunately, good old British drizzle means that heat stroke needn't worry most of us. Lucky old us.

Controlling Blood Sugar

Blood sugar is controlled as part of homeostasis, using the hormone insulin. Learn how it works.

Insulin Controls Blood Sugar Levels

1) Eating foods containing carbohydrate puts glucose into the blood from the gut.

2) Normal respiration (see p60) in cells removes glucose from the blood.

3) Vigorous exercise also removes a lot of glucose from the blood.

4) Levels of glucose in the blood must be kept steady. Changes in blood glucose are monitored and controlled by the pancreas, using insulin...

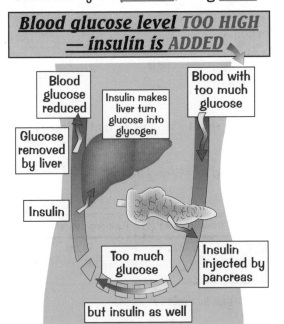

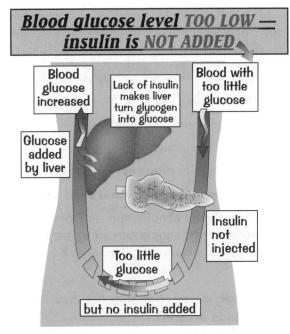

Glycogen can be stored in the liver until blood sugar levels get low again. Muscles have their own store.

Diabetes — the Pancreas Stops Making Enough Insulin

1) Diabetes (type 1) is a condition where the pancreas doesn't produce enough insulin.

2) The result is that a person's blood glucose level can rise to a level that can kill them.

3) The problem can be controlled in two ways:

a) Avoiding foods rich in simple carbohydrates, i.e. sugars (which cause glucose levels to rise rapidly). It can also be helpful to take exercise after eating to try and use up the extra glucose produced during digestion — but this isn't usually very practical.

b) Injecting insulin into the blood at mealtimes. This will make the liver remove the glucose as soon as it enters the blood from the gut, when the food is being digested. This stops the level of glucose in the blood from getting too high, and is a very effective treatment. However, the person must make sure they eat sensibly after injecting insulin, or their blood sugar could drop dangerously.

4) The amount of insulin that needs to be injected depends on the person's diet and how active they are.

5) Diabetics can check their blood sugar using a glucose-monitoring device. This is a little hand-held machine. They prick their finger to get a drop of blood for the machine to check.

My blood sugar feels low after all that — pass the biscuits...

This stuff can seem a bit confusing at first, but if you learn those two diagrams, it'll all start to get a lot easier. Don't forget that there are two ways to control diabetes — diet and injecting insulin.

Revision Summary for Section One

Congratulations, you've made it to the end of the first section. I reckon that section wasn't too bad, there's some pretty interesting stuff there — nerves, hormones, homeostasis... what more could you want? Actually, I know what more you could want... some questions to make sure you know it all.

1) List the five sense organs and the receptors that each one contains.

2) What do the letters CNS stand for? What does the CNS do? What does it consist of?

3) Where would you find the following receptors in a dog?

   a) smell   b) taste   c) light   d) pressure   e) sound.

4) What is the purpose of a reflex action?

5) Describe the pathway of a reflex arc from stimulus to response.

6) Draw a diagram of a typical neurone, labelling all its parts.

7) What's a synapse? How are signals passed across a synapse?

8) Describe the iris reflex. Why is this needed?

9) How does accommodation of the eye work? Is the lens fat or thin to look at distant objects?

10) Define the term 'hormone'.

11) Give two examples of hormones, saying where they're made and what they do.

12) List three differences between nerves and hormones.

13)* Here's a table of data about response times.
   a) Which response (if any) is carried by nerves?
   b) Which (if any) is carried by hormones?

Response	Reaction time (s)	Response duration (s)
A	0.005	0.05
B	2	10

14) What secondary sexual characteristics does testosterone trigger in males? And oestrogen in females?

15) Sketch a timeline of the 28-day menstrual cycle. Label the four stages of the cycle and show when the egg is released.

16) What roles do oestrogen, progesterone, FSH and LH play in the menstrual cycle?

17) State two advantages and two disadvantages of using the contraceptive pill.

18) Which hormone is used to stimulate egg production in fertility treatment?

19) Describe how IVF is carried out. Describe some of the main issues in the IVF debate.

20) What is meant by homeostasis?

21) Explain how negative feedback helps to maintain a constant internal environment.

22) Describe how the amount and concentration of urine you produce varies depending on how much exercise you do and how hot it is.

23) Describe how body temperature is reduced when you're too hot. What happens if you're cold?

24) Explain how insulin controls blood sugar levels.

25) Define diabetes and describe two ways in which it can be controlled.

* Answers on page 140

Eating Healthily

You might not think it but you don't just eat to stop you feeling <u>hungry</u> — you need a <u>balanced diet</u> to make sure that everything keeps working as it's supposed to.

A Balanced Diet Supplies All Your Essential Nutrients

A balanced diet gives you all the <u>essential nutrients</u> you need. The six essential nutrients are <u>carbohydrates</u>, <u>proteins</u>, <u>fats</u>, <u>vitamins</u>, <u>minerals</u> and <u>water</u>. You also need <u>fibre</u> (to keep the gut in good working order). Different nutrients are required for different functions in the body:

NUTRIENTS	FUNCTIONS
Carbohydrates	Carbohydrates (e.g. glucose) provide energy.
Fats	Fats provide energy, act as an energy store and provide insulation.
Proteins	Proteins are needed for growth and repair of tissue, and to provide energy in emergencies.
Vitamins	Various functions: e.g. vitamin C is needed to prevent scurvy.
Minerals	Various functions: e.g. iron is needed to make haemoglobin for healthy blood.
Water	We need a constant supply to replace water lost through urinating, breathing and sweating.

1) <u>Carbohydrates</u> are made up of <u>simple sugars</u> like <u>glucose</u>.

2) <u>Fats</u> are made up of <u>fatty acids</u> and <u>glycerol</u>.

3) <u>Proteins</u> are made up of <u>amino acids</u>.
Some amino acids can't be made by the body, so you have to get them from your <u>diet</u> — these are called <u>essential amino acids</u>. You can get all the essential amino acids by eating protein that comes from <u>animals</u> (in other words, meat). These animal proteins are called <u>first class proteins</u>. <u>Vegetarians</u> have to eat a <u>varied diet</u> to get all the essential amino acids they need.

Energy and Nutrient Needs Vary in Different People

A balanced diet <u>isn't</u> a set thing — it's <u>different</u> for everyone. The balance of the different nutrients a person needs depends on things like their <u>age</u>, <u>gender</u> and <u>activity level</u>.

Age	→	<u>Children</u> and <u>teenagers</u> need more <u>protein</u> for <u>growth</u>. <u>Older people</u> need more <u>calcium</u> to protect against <u>degenerative bone diseases</u> like <u>osteoporosis</u>.
Gender	→	<u>Females</u> need more <u>iron</u> to replace the iron <u>lost</u> in <u>menstrual blood</u>.
Physical activity	→	<u>Active people</u> need more <u>protein</u> for muscle development, and more <u>carbohydrate</u> for energy.

The amount of energy you need from your food also depends on your <u>resting metabolic rate</u>.

1) You need <u>energy</u> to fuel the chemical reactions in the body that keep you alive. These reactions are called your <u>metabolism</u>, and the speed at which they occur is your <u>metabolic rate</u>.

2) There are slight variations in the <u>resting metabolic rate</u> of different people. For example, <u>muscle</u> needs more energy than <u>fatty tissue</u>, which means (all other things being equal) people with a higher proportion of muscle to fat in their bodies will have a <u>higher</u> metabolic rate.

3) <u>Men</u> tend to have a slightly <u>higher</u> rate than <u>women</u> — they're generally slightly <u>bigger</u> and have a larger proportion of <u>muscle</u>. Other <u>genetic factors</u> may also have some effect.

4) And regular <u>exercise</u> can boost your resting metabolic rate because it <u>builds muscle</u>.

Diet tip — the harder you revise the more calories you burn...

Exercise is important too — people who <u>exercise regularly</u> are usually <u>fitter</u>. But being <u>fit</u> isn't the same as being <u>healthy</u> — you can be fit and slim, but still <u>unhealthy</u> because your diet isn't <u>balanced</u>.

Diet Problems

You are what you eat, apparently. That makes me baked beans. But at least I'm not toast.

In Developed Countries the Problem Is Often Too Much Food

In developed countries, obesity is becoming a serious problem. In the UK, 1 in 5 adults are obese, with obesity contributing to the deaths of over 30 000 people each year in England alone.

1) Obesity is defined as being 20% (or more) over your recommended body weight.

2) Too much sugary or fatty food and too little exercise are the main causes of obesity.

3) People can also be obese due to an underactive thyroid gland, but this problem isn't common.

4) Obesity can increase the risk of diabetes, arthritis, high blood pressure, coronary heart disease (CHD) and even some forms of cancer, e.g. breast cancer.

In Developing Countries There May Be Too Little Food

1) This can be a lack of one or more specific types of food, or not enough food of any sort (starvation). Young children, the elderly and women tend to suffer most and the effects vary depending on what foods are missing from the diet. Common problems include slow growth (in children), fatigue and poor resistance to infection.

2) Eating too little protein can cause a condition called kwashiorkor. A common symptom is a swollen stomach. Kwashiorkor is especially common in poorer developing countries — protein-rich foods are often too expensive to buy. Children need a greater proportion of protein than adults (for growth), so they may be more likely to suffer.

You can calculate the recommended daily allowance (RDA) of protein using the formula: $\text{RDA (g)} = 0.75 \times \text{body mass (kg)}$

3) Even in developed countries, some psychological disorders can cause under-nutrition, e.g. anorexia nervosa and bulimia nervosa. Anorexia nervosa leads to self-starvation. Bulimia nervosa involves bouts of binge eating, followed by self-induced vomiting. They're both usually caused by low self-esteem and anxiety about body fat — sufferers have a poor self-image. These disorders can cause a host of other illnesses, e.g. liver failure, kidney failure, heart attacks, muscle wastage, low blood pressure and mineral deficiencies. Both disorders can be fatal.

Body Mass Index Indicates If You're Under- or Overweight

The body mass index (BMI) is used as a guide to help decide whether someone is underweight, normal, overweight or obese. It's calculated from their height and weight:

$$\text{BMI} = \frac{\text{body mass}}{(\text{height})^2} \quad \begin{array}{l} \text{(in kg)} \\ \text{(in m)} \end{array}$$

The table shows how BMI is used to classify people's weight.

Body Mass Index	Weight Description
below 18.5	underweight
18.5 - 24.9	normal
25 - 29.9	overweight
30 - 40	moderately obese
above 40	severely obese

BMI isn't always reliable. Athletes have lots of muscle, which weighs more than fat, so they can come out with a high BMI even though they're not overweight. An alternative to BMI is measuring % body fat.

Too much or too little — it's a fine line to tread...

Studying problems like obesity and starvation is hard because accurate data can be difficult to collect. For example, starving people in developing countries may not reach medical aid. And with obesity the health problems tend to be more long-term, and people don't necessarily seek medical assistance.

Cholesterol and Salt

Aha, these two are <u>diet baddies</u>, right? Well, sort of. It's <u>more complicated</u> than you might think.

A <u>High Cholesterol Level</u> Is a <u>Risk Factor</u> for <u>Heart Disease</u>

1) <u>Cholesterol</u> is a <u>fatty substance</u> that's <u>essential</u> for good health. It's found in <u>every</u> cell in the body.

2) But you don't want too much of it because a <u>high cholesterol level</u> in the blood causes an <u>increased risk</u> of various problems — like coronary heart disease.

3) This is due to <u>blood vessels</u> getting <u>clogged</u> with fatty cholesterol deposits. This <u>reduces</u> blood flow to the heart, which can lead to <u>angina</u> (chest pain), or a <u>heart attack</u> (if the vessel is blocked completely).

4) The <u>liver</u> is really important in <u>controlling</u> the <u>amount</u> of cholesterol in the body. It <u>makes</u> new cholesterol and <u>removes</u> it from the blood so that it can be <u>eliminated</u> from the body.

5) The <u>amount</u> the liver makes depends on your <u>diet</u> (see below) and <u>inherited factors</u>.

Cholesterol is *Carried* <u>Around the Body</u> by <u>HDLs</u> and <u>LDLs</u>

1) Cholesterol is transported around the body in the <u>blood</u> by <u>lipoproteins</u> (i.e. fat attached to protein). These can be <u>high density lipoproteins</u> (<u>HDLs</u>), or <u>low density lipoproteins</u> (<u>LDLs</u>).

2) LDLs carry fat to the cells — they're called '<u>bad cholesterol</u>' as <u>excess</u> LDLs can cause a build up of cholesterol in the arteries.

3) HDLs carry cholesterol to the <u>liver</u> for removal from the body — they're called '<u>good cholesterol</u>'.

4) The <u>LDL/HDL balance</u> is important. Ideally, you want more HDLs than LDLs in the blood.

5) The level of cholesterol in the body is affected by <u>fat</u> in the <u>diet</u>. But it's not just the <u>amount</u> of fat, the <u>types</u> of fat you eat are even more crucial...

- <u>SATURATED FATS</u> <u>raise</u> blood cholesterol levels, so you should only eat them in moderation.

- <u>POLYUNSATURATED FATS</u> tend to <u>lower</u> blood cholesterol by increasing its removal from the body.

- <u>MONOUNSATURATED FATS</u> used to be considered 'neutral' for health. But recent evidence suggests they may help to <u>lower</u> blood cholesterol. People who have a diet high in monounsaturates tend to have <u>lower</u> levels of heart disease.

Too Much <u>Salt</u> *Can Cause* <u>High Blood Pressure</u>

1) Another <u>risk factor</u> (i.e. something that <u>increases the risk</u>) of heart disease is <u>high blood pressure</u> (<u>hypertension</u>).

2) Eating too much <u>salt</u> may cause <u>hypertension</u>. This is a problem for about 30% of the UK population who are '<u>salt sensitive</u>' and need to carefully monitor <u>how much salt they eat</u>.

3) This isn't always easy though — most of the salt you eat is probably in <u>processed foods</u> (such as breakfast cereals, soups, sauces, ready meals, biscuits...). The salt you <u>sprinkle</u> on food makes up quite a small proportion. And as if things weren't complicated enough... on food labels, <u>salt</u> is usually listed as <u>sodium</u>.

4) There are <u>other</u> risk factors for <u>high blood pressure</u> too — e.g. you're more likely to suffer from it as you get <u>older</u>, if you're <u>overweight</u>, if you drink too much <u>alcohol</u>, if you're <u>stressed</u>...

5) High blood pressure can lead to blood vessels <u>bursting</u>, which can cause all kinds of different problems depending on where in the body it happens — as well as heart attacks it can lead to <u>strokes</u>, <u>brain damage</u> and <u>kidney damage</u>. So it's a good idea to sort your lifestyle out.

My mother-in-law raises my blood pressure...

The more risk factors in your life, the <u>more likely</u> you are to suffer from a disease, but it doesn't mean you <u>definitely</u> will. A smoker with high cholesterol and high blood pressure is <u>30 times more likely</u> to develop heart disease than someone <u>without</u> these risk factors. But it's not guaranteed...

Health Claims

It's sometimes hard to figure out if health claims or adverts are true or not.

New Day, New Food Claim — It Can't All Be True

1) To get you to buy a product, advertisers aren't allowed to make claims that are untrue — that's illegal.

2) But they do sometimes make claims that could be misleading or difficult to prove (or disprove).

For example, some claims are just vague (calling a product "light" for instance — does that mean low calorie, low fat, something else...).

Alternatively, they might call a breakfast cereal "low fat", and that'd be true.
But that could suggest that other breakfast cereals are high in fat — when in fact they're not.

3) And every day there's a new food scare in the papers (eeek — we're all doomed).
Or a new miracle food (phew — we're all saved).

4) It's not easy to decide what to believe and what to ignore. But these things are worth looking for:

   a) Is the report a scientific study, published in a reputable journal?
   b) Was it written by a qualified person (not connected with the food producers)?
   c) Was the sample of people asked/tested large enough to give reliable results?
   d) Have there been other studies which showed similar results?

A "yes" to one or more of these is a good sign.

Not All Diets Are Scientifically Proven

With each new day comes a new celebrity-endorsed diet. It's a wonder anyone's overweight.

1) A common way to promote a new diet is to say, "Celebrity A has lost x pounds using it".

2) But effectiveness in one person doesn't mean much. Only a large survey can tell if a diet is more or less effective than just eating less and exercising more — and these aren't done often.

3) The Atkins diet was high profile, and controversial — so it got investigated. People on the diet certainly lost weight. But the diet's effect on general health (especially long-term health) has been questioned. The jury's still out.

4) Weight loss is a complex process. But just like with food claims, the best thing to do is look at the evidence in a scientific way.

It's the Same When You Look at Claims About Drugs

Claims about the effects of drugs (both medical and illegal ones) also need to be looked at critically. But at least here the evidence is usually based on scientific research.

STATINS
1) There's evidence that drugs called statins lower blood cholesterol and significantly lower the risk of heart disease in diabetic patients.

2) The original research was done by government scientists with no connection to the manufacturers. And the sample was big — 6000 patients.

3) It compared two groups of patients — those who had taken statins and those who hadn't. Other studies have since backed up these findings.

So control groups were used. And the results were reproducible.

But research findings are not always so clear cut...

CANNABIS
1) Many scientists have looked at whether cannabis use causes brain damage and mental health problems or leads to further drug taking. The results vary, and are sometimes open to different interpretations.

2) Basically, until more definite scientific evidence is found, no one's sure.

"Brad Pitt says it's great" is NOT scientific proof...

Learn what to look out for before you put too much faith in what you read. Then buy my book — 100% of the people I surveyed (i.e. both of them) said it had no negative effect whatsoever on their overall wellbeing!

Drugs

Drugs alter what goes on in your body. Your body's essentially a seething mass of <u>chemical reactions</u> — drugs can <u>interfere</u> with these reactions, sometimes for the better, sometimes not.

Drugs Can be Beneficial or Harmful, Legal or Illegal

1) Drugs are substances which <u>alter the chemical reactions in the body</u>.
   Some drugs are <u>medically useful</u>, such as <u>antibiotics</u> (e.g. <u>penicillin</u>).

2) But many drugs are <u>dangerous</u> if misused (this goes for both illegal drugs <u>and</u> legal ones).
   That could mean problems with either your <u>physical</u> or <u>mental</u> health.

3) This is why you can buy some drugs <u>over the counter</u> at a pharmacy, others are restricted so you can
   only get them on <u>prescription</u> (your <u>doctor decides</u> if you should have them), and others are <u>illegal</u>.

4) Some people get <u>addicted</u> to certain drugs — this means they have a physical
   need for that drug, and if they don't get it they get <u>withdrawal symptoms</u>. It's not
   just illegal drugs that are addictive — many legal ones are too, e.g. <u>caffeine</u>.
   Caffeine withdrawal symptoms include irritability and shaky hands.

5) <u>Tolerance</u> develops with some drugs — the body gets <u>used to having it</u> and so you need a <u>higher dose</u>
   to give the <u>same effect</u>. This can happen with legal drugs (e.g. alcohol), and illegal drugs (e.g. heroin).

6) Some drugs are <u>illegal</u> — usually because they're considered to be dangerous. In the UK, illegal drugs
   are classified into <u>three</u> main categories — <u>Classes A, B and C</u>. Which class a drug is in depends on
   how <u>dangerous</u> it is thought to be — Class A drugs are the most dangerous.

   - <u>CLASS A drugs</u> include heroin, LSD, ecstasy and cocaine.
   - <u>CLASS B drugs</u> include cannabis and amphetamines (speed). (Amphetamines are class A if prepared for injection.)
   - <u>CLASS C drugs</u> include anabolic steroids and tranquillisers.

Drugs Can Affect Your Behaviour

1) A lot of drugs affect your <u>nervous system</u>. Drugs can interfere with the way <u>signals</u> are sent around
   your body from <u>receptors</u> to the <u>brain</u>, and from the <u>brain</u> to <u>muscles</u> (see page 5).

2) The effects of drugs on the nervous system can alter <u>behaviour</u> (which has the potential to cause
   <u>danger</u> — either for the person who took the drug, or for others).

3) For example, <u>driving</u> and operating <u>machinery</u> aren't safe if you've taken certain drugs
   — e.g. alcohol, tranquillisers or cannabis (see page 21 for more info about alcohol).

4) Some drugs (e.g. alcohol) can also affect people's <u>judgement</u>. This could mean someone just
   'losing their inhibitions' — relaxing a bit at a party, for instance.

5) But it could mean they take more <u>risks</u> — e.g. <u>sharing needles</u> and having <u>unprotected sex</u> are more
   likely to happen under the influence of drink or drugs, increasing the risk of infections like <u>HIV</u>.

6) Drug abuse can also affect your <u>immune system</u> — making infections more <u>likely</u>.

Sedatives and Stimulants Have Opposite Effects on the Nervous System

- <u>Sedatives</u> (or <u>depressants</u>) — e.g. alcohol, barbiturates, solvents, temazepam.
  These <u>decrease</u> the <u>activity of the brain</u>. This slows down the <u>responses</u> of the <u>nervous
  system</u>, causing <u>slow reactions</u> and <u>poor judgement</u> of speed and distances.

- <u>Stimulants</u> — e.g. <u>nicotine, ecstasy, caffeine</u>.
  These do the opposite of depressants — they <u>increase</u> the <u>activity of the brain</u>.
  This makes you feel more <u>alert</u> and <u>awake</u>.

Drugs can kill you or cure you (or anything in between)...

Many people take <u>drugs of some kind</u>, e.g. caffeine, headache tablets, alcohol, hayfever medicine or an
inhaler for asthma. Most of these are okay if you're careful with them and don't go mad. It's <u>misuse</u>
that can get you into trouble (e.g. a paracetamol overdose can kill you). Read the packet.

Drug Testing

Drugs have <u>medical</u> uses too, obviously. But before they can be used, they have to be tested...

Medical Drugs Have to Be Thoroughly Tested

New drugs are constantly being <u>developed</u>. But before they can be given to the general public, they have to go through a <u>thorough</u> testing procedure. This is what usually happens...

1) <u>Computer models</u> are often used in the early stages — these simulate a human's response to a drug. This can identify promising drugs to be tested in the next stage (but sometimes it's not as accurate as actually seeing the effect on a <u>live organism</u>).

2) Drugs are then developed further by testing on <u>human tissues</u> in the lab. However, you can't use human tissue to test drugs that affect <u>whole</u> or <u>multiple</u> body systems, e.g. testing a drug for blood pressure must be done on a whole animal because you need an intact circulatory system.

3) The next step is to develop and test the drug using <u>live animals</u>. The law in Britain states that any new drug must be tested on <u>two</u> different <u>live mammals</u>. Some people think it's <u>cruel</u> to test on animals, but others believe this is the <u>safest</u> way to make sure a drug isn't dangerous before it's given to humans.

4) After the drug has been tested on animals it's tested on <u>human volunteers</u> in a <u>clinical trial</u> — this should determine whether there are any <u>side effects</u>.

Another argument is that animals are so different to humans that testing on them is often pointless.

<u>Clinical trials</u> usually work something like this...

1) There are <u>two groups</u> of patients. One is given the <u>new drug</u>, the other is given a <u>placebo</u> (a 'sugar pill' that looks like the real drug but doesn't do anything). This is done so scientists can see the actual difference the drug makes — it allows for the <u>placebo effect</u> (when the patient expects the treatment to work and so <u>feels better</u>, even though the treatment isn't doing anything).

2) Clinical trials are <u>blind</u> — the patient in the study <u>doesn't know</u> whether they're getting the drug or the placebo. In fact, they're often <u>double blind</u> — neither the <u>patient</u> nor the <u>scientist</u> knows until all the results have been gathered.

Developing New Drugs is Expensive

New drugs are often very sophisticated, and it can take <u>many years</u> to <u>develop</u> and <u>test</u> a drug to the stage where it can be put into use (and most potential drugs are <u>rejected</u> during the trials). So new drugs tend to cost a lot.

Things Have Gone Wrong in the Past

An example of what can happen when drugs are not thoroughly tested is the case of <u>thalidomide</u> — a drug developed in the 1950s.

1) Thalidomide was intended as a <u>sleeping pill</u>, and was tested for that use. But later it was also found to be effective in relieving <u>morning sickness</u> in pregnant women.

2) Unfortunately, thalidomide <u>hadn't</u> been <u>tested</u> as a drug for morning sickness, and so it wasn't known that it could pass through the placenta and affect the <u>foetus</u>, causing <u>stunted growth</u> of the foetus's arms and legs. In some cases, babies were born with no arms or legs at all.

3) About <u>10 000</u> babies were affected by thalidomide, and only about <u>half</u> of them survived.

4) The drug was <u>banned</u>, and more <u>rigorous</u> testing procedures were introduced.

5) Thalidomide has recently been reintroduced — as a treatment for <u>leprosy</u>, <u>AIDS</u> and certain <u>cancers</u>. But it <u>can't</u> be used in pregnant women.

A little learning is a dangerous thing...

Thalidomide was an attempt to <u>improve</u> people's lives which then caused some pretty tragic knock-on effects. Could the same thing happen <u>today</u>? Well, maybe not the exact same thing, but there's no such thing as <u>perfect</u> knowledge — you can never eliminate risk <u>completely</u>.

Smoking and Alcohol

Drugs are also used <u>recreationally</u>. Some of these are legal, others illegal. And some are more <u>harmful</u> than others. But two drugs that have a massive impact on people and society are both <u>legal</u>.

Smoking Tobacco Can Cause Quite a Few Problems

1) Tobacco smoke contains <u>carbon monoxide</u> — this <u>combines</u> irreversibly with <u>haemoglobin</u> in blood cells, meaning the blood can carry <u>less oxygen</u>. In pregnant women, this can deprive the <u>foetus</u> of oxygen, leading to the baby being born <u>underweight</u>.

2) Tobacco smoke also contains carcinogens — chemicals that can lead to <u>cancer</u>. Lung cancer is way more common among smokers than nonsmokers.

3) Disturbingly, the <u>incidence rate</u> (the number of people who get lung cancer) and the <u>mortality rate</u> (the number who die from it) aren't massively different — lung cancer kills <u>most</u> people who get it.

4) Smoking also causes <u>disease</u> of the <u>heart</u> and <u>blood vessels</u> (leading to <u>heart attacks</u> and <u>strokes</u>), and damage to the <u>lungs</u> (leading to diseases like <u>emphysema</u> and <u>bronchitis</u>).

5) And the <u>tar</u> in cigarettes damages the <u>cilia</u> (little hairs) in your lungs and windpipe (see p94). These hairs, along with <u>mucus</u>, catch a load of <u>dust</u> and <u>bacteria</u> before they reach the lungs. When these cilia are damaged, <u>chest infections</u> are more likely.

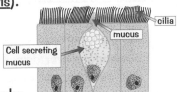

6) And to top it all off, smoking is <u>addictive</u> due to the <u>nicotine</u> in tobacco smoke.

Smoking and Lung Cancer Are Now Known to Be Linked

1) In the first half of the 20th century it was noticed that <u>lung cancer</u> and the popularity of <u>smoking</u> increased <u>together</u>. And studies found that far more <u>smokers</u> than <u>nonsmokers</u> got lung cancer.

2) But it was just a <u>statistical correlation</u> at that time (see p136) — it didn't <u>prove</u> that smoking <u>caused</u> lung cancer. Some people (especially in the tobacco industry) argued that there was some <u>other</u> factor (e.g. a person's <u>genes</u>) which both caused lung cancer, and also made people more likely to smoke.

3) Later research eventually <u>disproved</u> these claims. Now even the tobacco industry has had to admit that smoking does <u>increase</u> the <u>risk</u> of lung cancer.

Drinking Alcohol Can Do Its Share of Damage Too

1) The main effect of alcohol is to <u>reduce the activity</u> of the <u>nervous system</u> — slowing your reactions. It can also make you feel <u>less inhibited</u> — which can help people to socialise and relax with each other.

2) However, too much leads to <u>impaired judgement</u>, <u>poor balance</u> and <u>coordination</u>, <u>lack of self-control</u>, <u>unconsciousness</u> and even <u>coma</u>. Alcohol in excess also causes <u>dehydration</u>, which can damage <u>brain cells</u>, causing a noticeable <u>drop</u> in <u>brain function</u>. And too much drinking causes <u>severe damage</u> to the <u>liver</u>, leading to <u>liver disease</u>.

3) There are <u>social</u> costs too. Alcohol is linked with loads of murders, stabbings and domestic assaults.

These Two Legal Drugs Have a Massive Impact

1) <u>Alcohol</u> and <u>tobacco</u> have a bigger impact in the UK than illegal drugs, as <u>so many</u> people take them.

2) The National Health Service spends loads on treating people with <u>lung diseases</u> caused by <u>smoking</u> (or passive smoking). Add to this the cost to businesses of people missing days from work, and the figures get pretty scary.

3) The same goes for <u>alcohol</u>. The costs to the NHS are huge, but are pretty small compared to the costs related to <u>crime</u> (police time, damage to people/property) and the <u>economy</u> (lost working days etc.).

Drinking and smoking — it's so big and soooo clever...

So it's legal drugs that have the most impact on the country as a <u>whole</u> — when you take everything into consideration. Should the <u>Government</u> do more to try and reduce the number of people who smoke — or is it up to individual <u>people</u> what they do with their lives? No easy answer to that one...

Solvents and Painkillers

Two other groups of drugs you need to know about are solvents and painkillers.

Solvents Affect the Lungs and Neurones

1) Solvents are found in lighter fuel, spray paints, aerosols, thinners and dry cleaning fluids. They're useful chemicals, but can be misused as drugs (by inhaling the fumes).

2) Solvents act on the nervous system. Like alcohol, they're depressants — they slow down messages as they're passed along neurones (and can cause all sorts of other damage as well).

3) Solvent abuse often causes brain damage in the long term — this could show up as a personality change, sleeplessness or short-term memory loss, for example.

4) Most solvents also irritate the lungs and the breathing passages.

Paracetamol is a Painkiller

1) Paracetamol is a medicine that can relieve mild to moderate pain, and reduce fever.

2) Paracetamol is generally pretty safe, but an overdose can be deadly. Paracetamol overdose causes horrendous liver damage. If it isn't treated quickly (and I mean really quickly) it's very dangerous. And paracetamol's especially dangerous after alcohol, so it's not a good idea for hangovers.

3) A paracetamol overdose is particularly dangerous because the damage sometimes isn't apparent for 4-6 days after the drug's been taken. By that time, it's too late — there's nothing doctors can do to repair the damage. Dying from liver failure takes several days, and involves heavy-duty pain.

4) Paracetamol in normal doses won't damage the liver (though accidental overdoses are quite common).

Opiates and Cannabinoids are Used as Painkillers

Some types of painkillers can only be used under medical supervision.

Opiates
- Opiates include opium, morphine and codeine — they're all found in the opium poppy.
- Opiates are all painkillers. Morphine's used by doctors — it's very effective. But just like heroin, morphine's very addictive, and so it's illegal without a prescription.

Cannabis
- Cannabis has been used as a medicine for centuries, but it's now illegal.
- For years, no one really knew what cannabis did inside the body — this was because research on cannabinoids (the active ingredients in cannabis) was tricky (due to legal restrictions).
- The situation changed when scientists discovered receptors in the body for cannabinoids. Recent research seems to suggest that cannabinoids do provide benefits for some patients (though for most people, there's probably something better available).

Different Painkillers Work in Different Ways

1) Aspirin and ibuprofen work by inhibiting the formation of prostaglandins (chemicals which cause swelling, and sensitise the endings of nerves that register pain).

2) Paracetamol seems to work in a similar way to aspirin and ibuprofen, but scientists aren't really sure.

3) Opiates, like morphine and codeine, are very strong painkillers. They work by interfering with the mechanism by which 'pain-sensing' nerve cells transmit messages. They also act on the brain to stop it sensing the pain.

You can learn this — take the pain, take the pain...

Isn't it amazing that we're still not sure how paracetamol works... Apparently the pain-reducing effects of paracetamol were just discovered by accident. That's science for you — a series of accidents which add together to make amazing discoveries. Learn all the stuff, test yourself, and learn it again if need be.

Causes of Disease

An <u>infectious</u> disease is a disease that can be <u>transmitted</u> from one person to another — either <u>directly</u> (person to person), or <u>indirectly</u> (where some kind of <u>carrier</u> is involved, e.g. mosquitoes spread malaria, and certain bacteria are passed on in food or water). But obviously <u>not all</u> diseases are infectious.

Infectious Diseases are Caused by Pathogens

1) <u>Pathogens</u> are <u>microorganisms</u> (<u>microbes</u>) that cause <u>disease</u>.
2) They include some <u>bacteria</u>, <u>protozoa</u> (certain single-celled creatures), <u>fungi</u> and <u>viruses</u>.
3) All pathogens are <u>parasites</u> — they live off their host and give nothing in return.
4) Microorganisms can <u>reproduce very fast</u> inside a host organism.

Bacteria and Viruses are Very Different

...but they can both multiply quickly inside your body — they love the warm conditions.

1. Bacteria Are Very Small Living Cells

1) Bacteria are <u>very small cells</u> (about 1/100th the size of your body cells), which can reproduce rapidly inside your body.
2) They make you <u>feel ill</u> by doing <u>two</u> things:
   a) <u>damaging your cells</u>, b) <u>producing toxins</u> (poisons).
3) But... some bacteria are <u>useful</u> if they're in the <u>right place</u>, like in your digestive system.

TB (tuberculosis) is caused by bacteria.

Bacteria are cells with no nucleus. The DNA is free in the cytoplasm

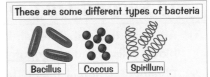

These are some different types of bacteria

Bacillus Coccus Spirillum

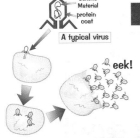

Genetic Material protein coat

A typical virus

eek!

2. Viruses Are Not Cells — They're Much Smaller

1) Viruses are <u>not cells</u>. They're <u>tiny</u>, about 1/100th the size of a bacterium. They're usually no more than a <u>coat of protein</u> around some genetic material.
2) They <u>replicate themselves</u> by invading the <u>nucleus</u> of a cell and using the <u>DNA</u> it contains to produce many <u>copies</u> of themselves. The cell will usually then <u>burst</u>, releasing all the new viruses.
3) This <u>cell damage</u> is what makes you feel ill.

Other Health Disorders Can be Caused in Various Ways

1) <u>Vitamin deficiency</u>, e.g. you can get <u>scurvy</u> if you don't get enough <u>vitamin C</u>.
2) <u>Mineral deficiency</u>, e.g. a lack of <u>iron</u> in the diet can lead to <u>anaemia</u>. Iron is needed to make the protein <u>haemoglobin</u> (which carries <u>oxygen</u> in the red blood cells).
3) <u>Genetic inheritance</u> of disorders (see p40), e.g. <u>red-green colour blindness</u> (sufferers find it hard to distinguish between red and green) and <u>haemophilia</u> (a blood clotting disorder).
4) <u>Body disorders</u> are caused by body cells not working properly, e.g. <u>diabetes</u> (see p13) and <u>cancer</u>.

Cancer is caused by body <u>cells growing out of control</u>

This forms a <u>tumour</u> (a mass of cells). Tumours can either be <u>benign</u> or <u>malignant</u>:

1) Benign — This is where the tumour grows until there's no more room. The cells <u>stay</u> where they are. This type <u>isn't</u> normally dangerous.

2) Malignant — This is where the tumour grows and can <u>spread</u> to other sites in the body. Malignant tumours are <u>dangerous</u> and can be fatal.

Smiling's infectious — but I dunno what the pathogen is...

A lot of microorganisms <u>won't</u> do you any harm — it's just the pathogens that you want to steer clear of — so don't eat rotten meat, etc. Your body's actually got loads of microorganisms inside it, all over it, everywhere. As has your house (even if it's just been cleaned). It's a fact of life — get used to it.

The Body's Defence Systems

Your body is constantly fighting off attack from all sorts of nasties — yep, things really are out to get you. The body has three lines of defence to stop things causing disease.

The First Line of Defence Stops Pathogens Entering the Body

The first line of defence consists mostly of physical barriers — they stop foreign bodies getting in.

1) The SKIN

Undamaged skin is a very effective barrier against microorganisms.

And if it gets damaged, the blood clots quickly to seal cuts and keep microorganisms out.

2) The RESPIRATORY SYSTEM

The nasal passage and trachea are lined with mucus and cilia which catch dust and bacteria before they reach the lungs.

cilia

mucus

goblet cell (secreting mucus)

nucleus

3) The EYES

Eyes produce (in tears) a chemical called lysozyme which kills bacteria on the surface of the eye.

This is a chemical barrier — not a physical one.

The Second Line of Defence is Non-Specific White Blood Cells

1) Anything that gets through the first line of defence and into the body should be picked up by white blood cells called phagocytes (a chemical barrier).

2) Phagocytes detect things that are 'foreign' to the body, e.g. microbes. They engulf microbes and digest them.

3) Phagocytes are non-specific — they attack anything that's not meant to be there.

microbes

White Blood Cell

4) The white blood cells also trigger an inflammatory response. Blood flow to the infected area is increased (making the area red and hot), and plasma leaks into the damaged tissue (which makes the area swell up) — this is all so that the right cells can get to the area to fight the infection.

The Third Line of Defence is Specific White Blood Cells

1. Some Produce Antibodies

1) Every invading cell has unique molecules (called antigens) on its surface.

2) When certain white blood cells come across a foreign antigen (i.e. one it doesn't recognise), they will start to produce proteins called antibodies to lock on to the invading cells and mark them out for destruction by other white blood cells. The antibodies produced are specific to that type of antigen — they won't lock on to any others.

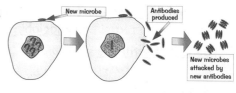

New microbe

Antibodies produced

New microbes attacked by new antibodies

3) Antibodies are then produced rapidly and flow round the body to mark all similar bacteria or viruses.

4) Some of these white blood cells stay around in the blood after the original infection has been fought off. They can reproduce very fast if the same antigen enters the body a second time. That's why you're immune to most diseases if you've already had them — the body carries a "memory" of what the antigen was like, and can quickly produce loads of antibodies if you get infected again.

2. Some Produce Antitoxins

These counter the toxins produced by invading microbes.

Cilia and mucus — biological self-defence...

The first line of defence stops nasty bugs getting in, while the other two fight infections once they're inside the body. It's amazingly clever, the human immune system — with all its different strategies to deal with invading organisms. But that's evolution for you, I guess...

Vaccinations

Vaccinations changed the way we deal with disease. Not bad for a little jab.

Immunisation — Protects from Future Infections

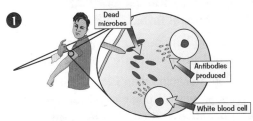

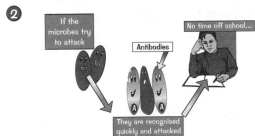

1) When you're infected with a new <u>microorganism</u>, it takes your white blood cells a few days to <u>learn</u> how to deal with it. But by that time, you can be pretty <u>ill</u>.

2) <u>Immunisation</u> involves injecting <u>dead</u> or <u>inactive</u> microorganisms. These carry <u>antigens</u>, which cause your body to produce <u>antibodies</u> to attack them — even though the microorganism is <u>harmless</u> (since it's dead or inactive). For example, the MMR vaccine contains <u>weakened</u> versions of the viruses that cause <u>measles</u>, <u>mumps</u> and <u>rubella</u> (German measles) stuck together.

3) But if live microorganisms of the same type appear after that, the white blood cells can <u>rapidly</u> mass-produce antibodies to help kill off the pathogen. Cool.

4) Vaccinations "wear off" over time. So <u>booster</u> injections can be given to increase levels of antibodies again.

Immunisation is classed as <u>active immunity</u>:

- <u>Active</u> immunity is where the immune system <u>makes its own antibodies</u> after being stimulated by a pathogen. It includes becoming <u>naturally immune</u> (see previous page) and <u>artificially immune</u> (immunisation). Active immunity is usually <u>permanent</u>.
- <u>Passive</u> immunity is where you use <u>antibodies made by another organism</u>, e.g. antibodies are passed from mother to baby through breast milk. Passive immunity is only <u>temporary</u>.

There are Benefits and Risks Associated with Immunisation

1) Immunisation <u>stops you from getting ill</u>... a pretty obvious benefit.

2) But there can be <u>short-term side effects</u>, e.g. <u>swelling</u> and <u>redness</u> at the site of injection and maybe feeling a bit <u>under the weather</u> for a week or two afterwards. And you can't have some vaccines if you're <u>already ill</u>, especially if your immune system is weakened.

3) But vaccinations mean we don't have to deal with a problem once it's happened — we can <u>prevent</u> it happening in the first place. Vaccines have helped <u>control</u> lots of infectious diseases that were once <u>common</u> in the UK (e.g. polio, measles, whooping cough, rubella, mumps, tetanus, TB...).

4) And if an outbreak does occur, vaccines can <u>slow down</u> or <u>stop</u> the spread (if people don't catch the disease, they won't pass it on).

5) Vaccination is now used all over the world. <u>Smallpox</u> no longer occurs at all, and <u>polio</u> infections have fallen by 99%.

6) However, some people think that immunisation can <u>cause other disorders</u>, e.g. one study <u>suggested</u> a link between the <u>MMR</u> (measles, mumps and rubella) vaccine and <u>autism</u>. Most scientists say the MMR jab is perfectly safe, but a lot of parents aren't willing to take the risk. This has led to a big rise in the number of children catching measles, and some people are now worried about an <u>epidemic</u>.

Prevention is better than cure...

Science isn't just about doing an experiment, finding the answer and telling everyone about it — scientists often disagree. Not that long ago different scientists had different opinions on the <u>MMR</u> vaccine — and argued about its safety. Many different studies were done before scientists concluded it was safe.

Treating Disease — Past and Future

The way we fight disease has <u>changed</u> loads over the last few decades. Thankfully.

Semmelweiss Cut Deaths by Using Antiseptics

1) While <u>Ignaz Semmelweiss</u> was working in Vienna General Hospital in the 1840s, he saw that women were dying in huge numbers after childbirth from a disease called puerperal fever.

2) He believed that <u>doctors</u> were spreading the disease on their <u>unwashed</u> hands. By telling doctors entering his ward to wash their hands in an <u>antiseptic solution</u>, he cut the death rate from 12% to 2%.

3) The antiseptic solution killed <u>bacteria</u> on doctors' hands, though Semmelweiss didn't know this (the <u>existence</u> of bacteria and their part in causing <u>disease</u> wasn't discovered for another 20 years). So Semmelweiss couldn't <u>prove</u> why his idea worked, and his methods were <u>dropped</u> when he left the hospital (letting death rates <u>rise</u> once again — d'oh).

4) Nowadays we know that <u>basic hygiene</u> is essential in controlling disease (though recent reports have found that a lack of it in some <u>modern</u> hospitals has helped the disease <u>MRSA</u> spread — see below).

Antibiotics Changed the Way We Fight Infections

1) <u>Antibiotics</u> were an incredibly important (but accidental) discovery. Some killer diseases (e.g. pneumonia and tuberculosis) suddenly became much easier to treat. The 1940s are sometimes called the era of the <u>antibiotics revolution</u> — it was that big a deal.

2) Unfortunately, bacteria <u>evolve</u> (adapt to their environment). If antibiotics are taken to deal with an infection but not all the bacteria are killed, those that survive may be slightly resistant to the antibiotic and go on to flourish. This process (an example of <u>natural selection</u>) eventually leaves you with an <u>antibiotic-resistant strain</u> of bacteria — not ideal.

3) A good example of antibiotic-resistant bacteria is <u>MRSA</u> (methicillin-resistant Staphylococcus aureus) — it's resistant to the powerful antibiotic methicillin. This is why it's important for patients to always <u>finish</u> a course of antibiotics, and for doctors to avoid <u>over-prescribing</u> them.

4) However, antibiotics <u>don't</u> destroy viruses. Viruses reproduce <u>using your own body cells</u>, which makes it very difficult to develop drugs that destroy just the virus without killing the body's cells.

5) <u>Flu</u> and <u>colds</u> are caused by <u>viruses</u>. Usually you just have to wait for your body to deal with the virus, and relieve the <u>symptoms</u> if you start to feel really grotty. There are some <u>antiviral</u> drugs available, but they're usually <u>reserved</u> for very <u>serious</u> viral illnesses (such as AIDS and hepatitis).

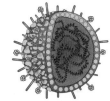

Flu virus. Lovely.

There Are New and Scary Dangers All the Time

1) For the last few decades, humans have been able to deal with <u>bacterial infections</u> pretty easily by using <u>antibiotics</u>.

2) But there'd be a big problem if a <u>virus</u> or bacterium evolved so that it was both <u>deadly</u> and could easily pass from <u>person</u> to <u>person</u>. (<u>Flu</u> viruses, for example, evolve quickly so this is quite possible.)

3) If this happened, <u>precautions</u> could be taken to stop the virus spreading in the first place (though this is hard nowadays — millions of people travel by plane every day). And <u>vaccines</u> and <u>antiviral</u> drugs could be developed (though these take <u>time</u> to mass produce).

4) But in the worst-case scenario, a flu <u>pandemic</u> (e.g. one evolved from bird flu) could kill billions of people all over the world.

A pandemic is when a disease spreads all over the world.

Antibiotic resistance is inevitable...

Antibiotic resistance is <u>scary</u>. Bacteria reproduce quickly, and so are pretty fast at <u>evolving</u> to deal with threats (e.g. antibiotics). If we were back in the situation where we had no way to treat bacterial infections, we'd have a <u>nightmare</u>. So do your bit, and finish your courses of antibiotics.

Revision Summary for Section Two

That was a long(ish) section, but kind of interesting, I reckon. These questions will show what you know and what you don't... if you get stuck, have a look back to remind yourself. But before the exam, make sure you can do all of them without any help — if you can't, you know you're probably <u>not ready</u>.

1) Name the six essential nutrients the body needs and say what each is used for.

2)* Put these people in order of how much energy they are likely to need from their food (from highest to lowest):   a) builder,   b) professional runner,   c) waitress,   d) secretary.

3) Define obesity and name three conditions obese people are at an increased risk of getting.

4) Explain what kwashiorkor is. Why is this condition more common in developing countries?

5) Explain what is meant by 'good cholesterol' and 'bad cholesterol'.

6) Why is it dangerous to have high levels of cholesterol?

7) Why is it bad for some people to eat too much salt?

8) If you don't add extra salt to your food, why will you not necessarily be safe from having too much salt in your diet?

9) Name four things that you could consider when trying to decide if a health claim is believable.

10) Explain the terms prescription drug and drug addiction.

11) How does a stimulant drug work? Give two examples.

12) Describe the four stages of drug testing.

13) What is a double blind clinical trial?

14) Name a drug that was not tested thoroughly enough and describe the consequences of its use.

15) Describe four different illnesses that smoking can cause.

16) How do carbon monoxide, carcinogens and tar in tobacco smoke each affect the body?

17) Alcohol is a depressant drug. Describe the symptoms of too much alcohol.
    What effect does alcohol have on the nervous system?

18)* Here is a graph of Mark's blood alcohol concentration against time.

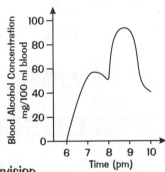

   a) When did Mark have his first alcoholic drink?

   b) When did Mark have his second alcoholic drink?

   c) The legal driving limit is 80 mg of alcohol per 100 ml of blood.
      Would Mark have been legally allowed to drive at 9pm?

19) Describe some of the effects that inhaling solvents can cause.

20) Why shouldn't you exceed the recommended dose of paracetamol?

21) Name two types of painkiller that can only be used under medical supervision.

22) How do aspirin and ibuprofen work? What about opiates?

23) Name the four types of microorganism that cause disease.

24) Explain the difference between benign and malignant tumours.

25) Name the three parts of the body which make up the first line of defence against pathogens.

26) What is the body's second line of defence against pathogens?

27) Explain how immunisation stops you getting infections.

28) Why shouldn't your doctor give you antibiotics for the flu?

* Answers on page 140

Section Two — Diet and Health

Adaptation

Animals and plants survive in many different <u>environments</u> — from <u>hot deserts</u> to <u>cold polar regions</u>, and pretty much everywhere in between. They can do this because they're <u>adapted</u> to their environment.

Desert Animals *Have Adapted to Save Water*

Animals that live in <u>hot</u>, <u>dry</u> conditions need to <u>keep cool</u> and use <u>water</u> efficiently.

LARGE SURFACE AREA COMPARED TO VOLUME	This lets desert animals <u>lose more body heat</u> — which helps to stop them overheating.

EFFICIENT WITH WATER
1) Desert animals <u>lose less water</u> by producing small amounts of <u>concentrated urine</u>.
2) They also make very little <u>sweat</u>. Camels are able to do this by tolerating <u>big changes</u> in <u>body temperature</u>, while kangaroo rats live in <u>burrows</u> underground where it's <u>cool</u>.

GOOD IN HOT, SANDY CONDITIONS
1) Desert animals have very thin layers of <u>body fat</u> to help them <u>lose</u> body heat. Camels keep nearly all their fat in their <u>humps</u>.
2) <u>Large feet</u> spread their <u>weight</u> across soft sand — making getting about easier.
3) A <u>sandy colour</u> gives good <u>camouflage</u> — so they're not as easy for their <u>predators</u> to spot.

Arctic Animals *Have Adapted to Reduce Heat Loss*

Animals that live in <u>really cold</u> conditions need to <u>keep warm</u>.

SMALL SURFACE AREA COMPARED TO VOLUME	Animals living in <u>cold</u> conditions have a <u>compact</u> (rounded) shape to keep their <u>surface area</u> to a minimum — this <u>reduces heat loss</u>.

WELL INSULATED
1) They also have a thick layer of <u>blubber</u> for <u>insulation</u> — this also acts as an <u>energy store</u> when food is scarce.
2) <u>Thick hairy coats</u> keep body heat in, and <u>greasy fur</u> sheds water (this <u>prevents cooling</u> due to evaporation).

GOOD IN SNOWY CONDITIONS
1) Arctic animals have <u>white fur</u> to match their surroundings — for <u>camouflage</u>.
2) <u>Big feet</u> help by <u>spreading weight</u> — which stops animals sinking into the snow or breaking thin ice.

Some Plants *Have Adapted to Living in a Desert*

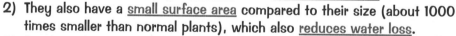

Desert-dwelling plants make best use of what little water is available.

MINIMISING WATER LOSS
1) Cacti have <u>spines instead of leaves</u> — to <u>reduce water loss</u>.
2) They also have a <u>small surface area</u> compared to their size (about 1000 times smaller than normal plants), which also <u>reduces water loss</u>.
3) A cactus <u>stores water</u> in its thick stem.

MAXIMISING WATER ABSORPTION
Some cacti have <u>shallow</u> but <u>extensive roots</u> to <u>absorb</u> water quickly over a large area. Others have <u>deep roots</u> to access <u>underground water</u>.

Some Plants *and Animals Are Adapted to Deter Predators*

There are various <u>special features</u> used by animals and plants to help <u>protect</u> them against being <u>eaten</u>.
1) Some plants and animals have <u>armour</u> — like roses (with <u>thorns</u>), cacti (with <u>sharp spines</u>) and tortoises (with <u>hard shells</u>).
2) Others produce <u>poisons</u> — like bees and poison ivy.
3) And some have amazing <u>warning colours</u> to scare off predators — like wasps.

Cactus spines — nasty.

In a nutshell, it's horses for courses...

It's <u>no accident</u> that animals and plants look like they do. So by looking at an animal's <u>characteristics</u>, you should be able to have a pretty good guess at the kind of <u>environment</u> it lives in — or vice versa. Why does it have a large/small surface area... what are those spines for... why is it white... and so on.

Classification

Scientists <u>classify</u> species so that anyone who reads their research knows exactly <u>which organism</u> it's about.

Classification *is* <u>Organising</u> Living Organisms *into* Groups

1) Nowadays scientists classify organisms into groups based on <u>genetic similarities</u>. For example, bats, whales and humans might <u>seem</u> quite different, but they have a similar bone pattern in their forelimbs, and they're all <u>genetically</u> related. The classification system reflects these <u>similarities</u> (they're all mammals).

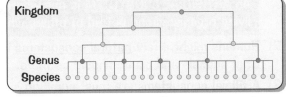

2) Living things are divided into <u>kingdoms</u> (e.g. the animal kingdom, the plant kingdom, etc.). Kingdoms are then <u>subdivided</u> into smaller and smaller groups. An example of one of these smaller groups is a <u>genus</u>.

3) A <u>genus</u> is a group of closely-related <u>species</u> — and a species is a group of <u>closely-related</u> organisms that can breed to produce <u>fertile offspring</u> (see next page).

4) The <u>binomial system</u> that's used to name organisms uses the Latin names of the genus and the species they belong to. E.g. humans are <u>Homo sapiens</u> — 'Homo' is our <u>genus</u> name and 'sapiens' is our <u>species</u>.

Living Things *Can be* Plants, Animals *or* Something Else

1) To be a member of the plant kingdom, organisms must contain <u>chloroplasts</u> and therefore be able to <u>make their own food</u> by photosynthesis (see page 78).

2) Members of the <u>animal</u> kingdom move about from place to place and have <u>compact</u> bodies (unlike plants, which spread out to catch as much light and water as possible and can't move about freely). Animals <u>can't make their own food</u> so they have to find things to eat, such as plants or other animals.

3) Other organisms, like <u>fungi</u> and <u>bacteria</u>, don't have animal or plant features and are put in <u>other kingdoms</u>.

4) Some single-celled organisms have features of <u>both</u> plants and animals. <u>Euglena</u> can <u>move</u> from place to place by thrashing its <u>flagellum</u>, but also has <u>chloroplasts</u> which allow it to make its own food. It's put into a kingdom called <u>Protoctista</u>, along with some other single-celled organisms.

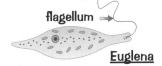

flagellum

Euglena

Vertebrates *Have Backbones*

The animal kingdom is divided into <u>vertebrates</u> and <u>invertebrates</u>. Vertebrates are animals with a <u>backbone</u> and an <u>internal skeleton</u>. Invertebrates don't have these structures — some do have an <u>external skeleton</u> though. <u>Vertebrates</u> are divided into five groups, called <u>classes</u> — fish, amphibians, reptiles, birds and mammals.

1) <u>FISH</u> live in water. They have <u>scales</u>, and <u>gills</u> for gas exchange.

2) <u>AMPHIBIANS</u> exchange gas partly through their skin, so gases must be able to move in and out — their skin's got to be <u>permeable</u> and <u>moist</u>.

3) <u>REPTILES</u> are more adapted to live on the land. They've got a <u>dry scaly skin</u> which stops them losing too much water.

4) Most <u>BIRDS</u> can fly and they've got <u>feathers</u> to help them do this. You'll also find a <u>beak</u> — useful for cracking seeds or catching prey.

5) <u>MAMMALS</u> have <u>fur</u> covering their bodies to keep them warm. They give birth to their young (rather than laying eggs like other vertebrates) and <u>produce milk</u> to feed them.

The rules of the classification system were made up using the animals and plants that were known about at the time. Sometimes <u>newly discovered species</u> don't really fit into any of the categories. These can be <u>living species</u> or <u>fossil ones</u>, such as <u>archaeopteryx</u>, which had reptilian teeth, clawed hands and a long bony tail, like a dinosaur, but also had wings and flight feathers, like a bird.

I'm not a vertebrate — I'm completely spineless...

There are <u>loads</u> of different types of organisms out there — so no wonder the classification system gets a bit unwieldy. This makes life no easier for you, I'm afraid — you've just got to <u>learn</u> it...

Ecosystems and Species

An ecosystem is all the different organisms living together in a particular environment. Sounds cosy.

Artificial Ecosystems Can be Carefully Controlled

1) There are two types of ecosystem you need to know about:

> A NATURAL ECOSYSTEM is one where humans don't control the processes going on within it.
> An ARTIFICIAL ECOSYSTEM is one where humans deliberately promote the growth of certain living organisms and get rid of others which threaten their well-being.

2) Humans might affect natural ecosystems in some way, but they don't take deliberate steps to decide what animals and plants should be there.

3) Artificial ecosystems are most common in money-making enterprises, e.g. farms and market gardens. Things like weedkillers, pesticides and fertilisers are used to control conditions. Artificial ecosystems normally have a smaller number of species (less biodiversity) than natural ones.

Estimate Population Sizes in an Ecosystem Using a Quadrat

A quadrat is a square frame enclosing a known area. You just place it on the ground, and look at what's inside it. To estimate population size:

1) Count all the organisms in a 1 m² quadrat.

2) Multiply the number of organisms by the total area (in m²) of the habitat.

3) Er, that's it. Bob's your uncle.

A quadrat

Two Important Points About This Kind of Counting Method...

1) The sample size affects the accuracy of the estimate — the bigger your sample, the more accurate your estimate of the total population is likely to be. (So it'd be better to use the quadrat more than once, get an average value for the number of organisms in a 1 m² quadrat, then multiply that by the total area.)

2) The sample may not be representative of the population, i.e. what you find in your particular sample might be different from what you'd have found if you'd looked at other bits of the habitat.

Different Organisms Belong to Different Species

1) Organisms are of the same species if they can breed to produce fertile offspring.

2) If you interbreed a male from one species with a female from a different species you'll get a hybrid (that's if you get anything at all). For example, a mule is a cross between a donkey and a horse. But hybrids are infertile so they aren't new species.

Unrelated Species May Have Similar Features

1) Similar species often share a recent common ancestor, so they're closely related in evolutionary terms. They often look very alike and tend to live in similar types of habitat, e.g. whales and dolphins.

2) This isn't always the case though — closely related species may look very different if they have evolved to live in different habitats, e.g. llamas and camels.

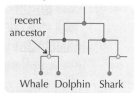

recent ancestor

Whale Dolphin Shark

3) Species that are very different genetically may also end up looking alike. E.g. dolphins and sharks look pretty similar and swim in a similar way. But they're totally different species — dolphins are mammals and sharks are fish.

4) So to explain the similarities and differences between species, you have to consider how they're related in evolutionary terms AND the type of environment they've adapted to survive in.

Ecosystems — aren't they used in submarines...

It's possible to breed lions and tigers together... it's true — they produce hybrids called tigons and ligers. They look a bit like lions and a bit like tigers... as you'd expect. In the same way, a bat is (I think) just a hybrid of a bird and a cat. And a donkey is the result of breeding a dog and a monkey.

Populations and Competition

Organisms have to <u>compete for resources</u> in the environment where they live.

Population Size <u>is Limited by</u> Available Resources

<u>Population size</u> is limited by:

1) The <u>total amount of food</u> or nutrients available
   (plants don't eat, but they get <u>minerals</u> from the soil).

2) The amount of <u>water</u> available.

3) The <u>amount of light available</u> (this applies only to plants really).

4) The quality and amount of <u>shelter</u> available.

Animals and plants of the <u>same</u> species and of <u>different</u> species will <u>COMPETE</u>
against each other for these resources. They all want to <u>survive</u> and <u>reproduce</u>.

<u>Similar organisms</u> will be in the <u>closest competition</u> — they'll be competing for the same <u>ecological niche</u>.

> <u>Example: Red and Grey Squirrels</u>
> 1) These two different species like the same kind of <u>habitat</u>, same kind of <u>food</u>, type of <u>shelter</u>, etc.
> 2) Grey squirrels are <u>better adapted</u> to <u>deciduous woodland</u>, so when they were introduced into Britain,
>   red squirrels <u>disappeared</u> from many areas — they just couldn't compete.

Populations of <u>Prey</u> and Predators <u>Go in Cycles</u>

In a community containing <u>prey</u> and <u>predators</u> (as most of them do of course):

1) The <u>population</u> of any species is usually <u>limited</u> by the amount of <u>food</u> available.

2) If the population of the <u>prey</u> increases, then so will the population of the <u>predators</u>.

3) However, as the population of predators <u>increases</u>, the number of prey will <u>decrease</u>.

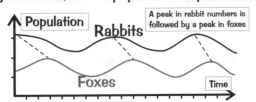

A peak in rabbit numbers is followed by a peak in foxes

e.g. <u>More grass</u> means <u>more rabbits</u>.
More rabbits means <u>more foxes</u>.
But more foxes means <u>less rabbits</u>.
Eventually less rabbits will mean <u>less foxes again</u>.
This <u>up and down pattern</u> continues...

Parasites <u>and</u> Mutualistic Relationships

The <u>survival</u> of some organisms can <u>depend</u> almost entirely on the presence of <u>other species</u>.

1) <u>PARASITES</u> live off a host. They <u>take</u> what they need to survive, <u>without</u> giving anything <u>back</u>.
   This often <u>harms</u> the host — which makes it a win-lose situation.

> • <u>Tapeworms</u> absorb lots of <u>nutrients</u> from the host, causing them to suffer from <u>malnutrition</u>.
> • <u>Fleas</u> are parasites. Dogs gain nothing from having fleas (unless you count hundreds of bites).

2) <u>MUTUALISM</u> is a relationship where <u>both</u> organisms benefit — so it's a win-win relationship.

> • Most plants have to rely on <u>nitrogen-fixing bacteria</u> in the soil to get the <u>nitrates</u> that they
>   need. But <u>leguminous plants</u> carry the bacteria in <u>nodules</u> in their <u>roots</u>. The bacteria get a
>   constant supply of <u>sugar</u> from the plant, and the plant gets essential <u>nitrates</u> from the bacteria.
> • 'Cleaner species' are fantastic. E.g. <u>oxpeckers</u> live on the backs of <u>buffalo</u>. Not only do they
>   <u>eat pests</u> on the buffalo, like ticks, flies and maggots (providing the oxpeckers with a source
>   of food), but they also <u>alert</u> the animal to any <u>predators</u> that are near, by hissing.

Revision stress — don't let it eat you up...

In the exam you might get asked about the distribution of <u>any</u> animals or plants. Just think about what
the organisms would need to survive. And remember, if things are in <u>limited supply</u> then there's going
to be <u>competition</u>. And the more similar the needs of the organisms, the more they'll have to compete.

Evolution

We've identified about <u>1.5 million different species</u> and there's loads more. So how did they all get here...?

No One Knows How Life Began

We know that <u>living things</u> come from <u>other</u> living things — that's easy enough.
But where the <u>first</u> living thing came from... that's a much more difficult question.

1) There are various <u>theories</u> suggesting how life first came into being, but no one really <u>knows</u>.

2) Maybe the first life forms came into existence in a primordial <u>swamp</u> (or under the <u>sea</u>) here on <u>Earth</u>. Maybe simple organic molecules were brought to Earth on <u>comets</u> — these could have then become more <u>complex</u> organic molecules, and eventually very simple <u>life forms</u>.

3) These ideas (and others) have their supporters. But we just don't know — the evidence was lost long ago. All we know is that life started <u>somehow</u>. And after that, we're on slightly firmer ground...

The Fossil Record Shows That Organisms Have Evolved

1) A <u>fossil</u> is <u>any evidence</u> of an animal or plant that lived ages ago.

2) Fossils form in rocks as <u>minerals</u> replace <u>slowly decaying</u> tissue (or in places where no decay happens) and show features like <u>shells</u>, <u>skeletons</u>, <u>soft tissue</u> (occasionally), <u>footprints</u>, etc. They show what was on Earth millions of years ago. They can also give clues about an organism's <u>habitat</u> and <u>food</u>.

3) We also know that the <u>layers of rock</u> where fossils are found were made at <u>different times</u>. This means it's possible to tell how long ago a particular species <u>lived</u>.

4) From studying the <u>similarities</u> and <u>differences</u> between fossils in rocks of different ages, we can see how species have <u>evolved</u> (changed and developed) over <u>billions of years</u>.

> **THEORY OF EVOLUTION:** Life began as simple organisms from which more complex organisms evolved (rather than just popping into existence).

6) In theory, you could put all species on a 'family tree' — where each new branch shows the <u>evolution</u> of a new species. Then you could easily find the most recent <u>common ancestor</u> of any two species. The more <u>recent</u> the common ancestor, the more <u>closely related</u> the two species.

5) Unfortunately, very few organisms <u>turn into fossils</u> when they die — most decay away completely. This leaves <u>gaps</u> in the <u>fossil record</u>, which means there are many species that we'll never know about.

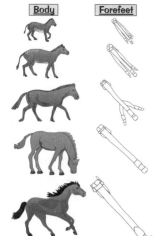

Suggested evolution of the horse
Body   Forefeet

The Evolution of the Horse

1) The fossil record of the <u>horse</u> provides <u>strong evidence</u> for the <u>theory of evolution</u>, but things are a little more complicated than we first thought.

2) If you stick all the fossil bones in order of age, they seem to show the modern horse evolving gradually from a creature about the size of a <u>dog</u>, with the <u>middle toe</u> slowly getting bigger to form the familiar <u>hoof</u>.

3) At first some fossils didn't seem to fit. But now we know that <u>several now-extinct kinds</u> of horse evolved at the same time, and it all makes <u>sense</u>.

There are Other Views About the Fossil Record

Some people interpret the fossil evidence differently. E.g. <u>creationists</u> believe that <u>each species</u> was created separately by God and will <u>never evolve</u> into new species. They don't think the fossil record is evidence for <u>gradual evolution</u>, but simply shows <u>a lot of different organisms</u>, some of which are now extinct.

Cell, blob, toad, monkey, me — what a fine family tree...

The fossil record provides good <u>evidence</u> for evolution, but it <u>can't prove it</u>. But proving a theory of something that happens over millions of years was never going to be straightforward, I guess.

Evolution

The <u>theory of evolution</u> (see last page) states that one of your (probably very distant) ancestors was a <u>blob</u> in a swamp somewhere. Something like that, anyway.

Make Sure You Know the Theory of Evolution

1) Don't forget what you learned on the last page — the theory of evolution states that all the animals and plants on Earth gradually '<u>evolved</u>' over <u>millions</u> of years, rather than just suddenly appearing.

2) Life on Earth began as <u>simple organisms</u> from which all the more complex organisms evolved. And it only took about <u>3 000 000 000 years</u>.

There are Lots of Modern Examples of Evolution

1) Peppered Moths Adapted Their Colour

<u>Peppered moths</u> are often seen on the <u>bark</u> of trees. Until the 19th century, the only ones found in England were <u>light</u> in colour. Then some areas became <u>polluted</u> and the soot darkened the tree trunks. A <u>black</u> variety of moth was found. The moths had <u>adapted</u> to stay <u>camouflaged</u>.

2) Bacteria Adapt to Beat Antibiotics

The "<u>survival of the fittest</u>" (see next page) affects bacteria just the same as other living things. They adapt to become <u>resistant</u> to our bacteria-fighting weapons — <u>antibiotics</u>.

1) If someone gets ill they might be given an <u>antibiotic</u> which <u>kills</u> 99% of the bacteria.

2) The 1% that survive are <u>resistant</u> so if they're passed on to somebody else, the antibiotic won't help them. Nowadays bacteria are getting resistant at such a rate, the development of antibiotics <u>can't keep up</u>. Eeek!

3) Rats Adapt to Beat Poison

The poison <u>warfarin</u> was widely used to control the <u>rat</u> population. However, a certain gene gives rats <u>resistance</u> to it, so rats which carry it are more likely to survive and breed. This gene has become more and more <u>frequent</u> in the rat population, so warfarin isn't as much use any more.

Environmental Change Can Cause Extinction

1) If a species can't evolve <u>quickly enough</u>, it's in trouble.

2) The <u>dinosaurs</u> and <u>woolly mammoths</u> became <u>extinct</u>, and it's only <u>fossils</u> that tell us they ever existed.

There are <u>three ways</u> a species can become <u>extinct</u>:
1) The <u>environment changes</u> more quickly than the species can adapt.
2) A new <u>predator</u> or <u>disease</u> kills them all.
3) They can't <u>compete</u> with another (new) species for <u>food</u>.

3) As the environment changes, it'll <u>gradually</u> favour certain characteristics (see next page).

4) Over many generations those features will be present in <u>more</u> of the population. In this way, the species constantly <u>adapts</u> to its changing environment.

5) But if the environment changes too <u>fast</u> the whole species may become <u>extinct</u>.

Did you know exams evolved from the Spanish Inquisition...

...well that's not really true. But it is true that humans and apes both evolved from a <u>common ancestor</u>. Scientists reckon that 5-8 million years ago the species <u>separated</u> and one population evolved into humans and the other into apes — so you're a distant relation (250 000th cousin, maybe) of a chimp.

Natural Selection

Charles Darwin developed a theory about how <u>evolution</u> actually happened. He called it the theory of <u>natural selection</u>. This is how he came up with it...

Darwin Made Four Important Observations...

1) All organisms produce <u>more offspring</u> than could possibly survive (e.g. only a few frogspawn survive and become frogs).

2) But in fact, population numbers tend to remain <u>fairly constant</u> over long periods of time.

3) Also, organisms in a species show <u>wide variation</u> in <u>characteristics</u>.

4) <u>Some</u> of the variations are <u>inherited</u>, and so <u>passed on</u> to the next generation.

...and Then Made These Two Deductions:

1) Since most offspring don't survive, all organisms must have to <u>struggle for survival</u>. <u>Being eaten</u>, <u>disease</u> and <u>competition</u> cause large numbers of individuals to die.

2) The ones who have characteristics that allow them to <u>survive and reproduce</u> better (i.e. the most useful <u>adaptations</u> to the environment) will <u>pass on these characteristics</u>.

This is the famous "<u>survival of the fittest</u>" statement. Organisms with slightly less survival value will probably perish first, leaving the <u>fittest</u> to pass on their <u>genes</u> to the next generation.

Organisms with Certain Characteristics Will Survive Better

Here's an example... once upon a time maybe all rabbits had <u>short ears</u> and managed OK. Then one day out popped a rabbit with <u>big ears</u> who could hear better and was always the first to dive for cover at the sound of a predator. Pretty soon he's fathered a whole family of rabbits with <u>big ears</u>, all diving for cover before the other rabbits, and before you know it there are only <u>big-eared</u> rabbits left — because the rest just didn't hear trouble coming quick enough.

This is how populations <u>adapt</u> to changes in their environment (an organism doesn't actually change when it's alive — changes only occur from generation to generation).

FOX!

Over many generations the <u>characteristic</u> that <u>increased survival</u> becomes <u>more common</u> in the population. If members of a species are separated somehow, and evolve in different ways to adapt to different conditions, then over time you can end up with totally <u>different species</u>.

Darwin's Theory Wasn't Popular at First

1) Darwin's theory <u>caused some trouble</u> at the time — it was the first plausible explanation for our own existence <u>without</u> the need for a "Creator". This was <u>bad news</u> for the religious authorities of the time, who tried to ridicule old Charlie's ideas. The idea that humans and monkeys had a common ancestor was hard for people to accept, and easy to take the mick out of.

2) Some <u>scientists weren't keen</u> either, at first. Darwin didn't provide a proper explanation of exactly <u>how</u> individual organisms passed on their survival characteristics to their offspring.

3) Later, the idea of <u>genetics</u> was understood — which <u>did</u> explain how characteristics are inherited.

Natural selection... sounds like vegan chocolates...

This is a good example of how scientific theories come about — someone <u>observes</u> something and then tries to <u>explain</u> it. Their theory will then be <u>tested</u> by other scientists using <u>evidence</u> — if the theory passes these tests, it gains in credibility. If not, it's <u>rejected</u>. Natural selection <u>hasn't</u> been rejected yet.

Revision Summary for Section Three

There's a lot to remember in this section and you need to know all the facts... so here are some questions to help you. If you get any wrong, there's no shame in it and nobody will shout at you — just go back and learn the stuff again AND DO IT PROPERLY THIS TIME.

1) Give four ways in which a desert animal may be adapted to its environment, and two ways that a desert plant might be adapted.
2) Explain how an animal that lives in the Arctic might be adapted to its environment.
3) State three ways that plants and animals might be adapted to deter predators.
4) Name two different kingdoms.
5) What do all vertebrates have in common?
6) Name the five different types of vertebrate.
7) Why are euglena and archaeopteryx difficult to classify?
8) What is the difference between a natural and an artificial ecosystem?
9) How could you estimate a population size in a habitat using a quadrat? Give two reasons why your results might not be 100% accurate.
10) Explain two reasons why different species may look similar.
11) Name three things that:
    a) plants compete for,   b) animals compete for.
12) Sketch a typical graph of prey and predator populations and explain the pattern shown.
13) What is the difference between a parasitic and a mutualistic relationship? Give an example of each.
14) Briefly describe two theories that have been suggested as explanations of how life on Earth began.
15) Write down the theory of Evolution.
16) Why are there gaps in the fossil record?
17) Describe three modern examples of evolution.
18) Give three reasons why some species become extinct.
19) What were Darwin's four observations and two deductions that led to his theory of natural selection?
20) Why was Darwin's theory controversial?

Variation in Plants and Animals

The word 'variation' sounds far too fancy for its own good. All it means is how animals or plants of the same species look or behave slightly differently from each other. You know, a bit taller or a bit fatter or a bit more scary-to-look-at etc. There are two kinds of variation — genetic and environmental.

Genetic Variation is Caused by Genes (Surprise)

1) All animals (including humans) are bound to be slightly different from each other because their genes are slightly different.

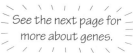
See the next page for more about genes.

2) Genes are the code which determines how your body turns out — they control your inherited traits, e.g. eye colour. We all end up with a slightly different set of genes. The exceptions to this rule are identical twins, because their genes are exactly the same.

Most Variation in Animals is Due to Genes and Environment

1) Most variation in animals is caused by a mixture of genetic and environmental factors.

2) Almost every single aspect of a human (or other animal) is affected by our environment in some way, however small. In fact it's a lot easier to list the factors which aren't affected in any way by environment:

> If you're not sure what "environment" means, think of it as "upbringing" instead.

1) Eye colour,
2) Hair colour in most animals (in humans, vanity plays a big part),
3) Inherited disorders like haemophilia, cystic fibrosis, etc.,
4) Blood group.

3) Environment can have a large effect on human growth even before someone's born. For example, a baby's weight at birth can be affected by the mother's diet.

4) And having a poor diet whilst you're growing up can stunt your growth — another environmental variation.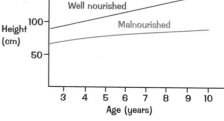

Growth Chart
Well nourished
Malnourished
Height (cm) 150 — 100 — 50
Age (years) 3 4 5 6 7 8 9 10

5) For some characteristics, it's hard to say which factor is more important — genes or environment...

   i) Health — Some people are more likely to get certain diseases (e.g. cancer and heart disease) because of their genes. But lifestyle also affects the risk, e.g. if you smoke or only eat junk food.

   ii) Intelligence — One theory is that although your maximum possible IQ might be determined by your genes, whether you get to it depends on your environment, e.g. your upbringing and school life.

   iii) Sporting ability — Again, genes probably determine your potential, but training is important too.

Environmental Variation in Plants is Much Greater

Plants are strongly affected by:
1) sunlight, 2) moisture level, 3) temperature, 4) the mineral content of the soil.

For example, plants may grow twice as big or twice as fast due to fairly modest changes in environment such as the amount of sunlight or rainfall they're getting, or how warm it is or what the soil is like.

> Think about it — if you give your pot plant some plant food (full of lovely minerals), then your plant grows loads faster. Farmers and gardeners use mineral fertilisers to improve crop yields.

Environmental variation — like sun and scattered showers...

So there you go... the "nature versus nurture" debate (*Are you like you are because of the genes you're born with, or because of the way you're brought up?*) summarised in one page. And the winner is... well, both of them really. Your genes are pretty vital, but then so is your environment. What an anticlimax.

Genes and Chromosomes

This page is a bit tricky, but it's dead important you get to grips with all the stuff on it — because you're going to hear a lot more about it over the next few pages...

1) Most cells in your body have a <u>nucleus</u> — and it's the nucleus that contains your <u>genetic material</u>.

nucleus

2) The human cell nucleus contains <u>23 pairs of chromosomes</u>. They are all well known and numbered. We all have two No. 19 chromosomes and two No. 12s etc.

A single <u>chromosome</u>.

A <u>pair</u> of <u>chromosomes</u>. (They're always in pairs, one from each <u>parent</u>.)

3) Chromosomes carry <u>genes</u>. Different genes <u>control</u> the development of different <u>characteristics</u>, e.g. hair colour.

DNA molecule

4) A <u>gene</u> — a <u>short length</u> of the chromosome...

...which is quite a long length of <u>DNA</u>.

The arms are held together in the centre.

Genes can exist in <u>different versions</u>. Each version gives a <u>different</u> variation of a <u>characteristic</u>, like blue or brown eyes. The different versions of the same gene are called <u>alleles</u> instead of genes — it's more sensible than it sounds!

5) The DNA is <u>coiled up</u> to form the <u>arms</u> of the <u>chromosome</u>.

It's hard being a DNA molecule, there's so much to remember...

This is the nuts and bolts of genetics, so you definitely need to understand <u>everything</u> on this page or you'll find the rest of this topic dead hard. The best way to get all of these important facts engraved in your mind is to <u>cover</u> the page, <u>scribble</u> down the main points and <u>sketch</u> out the diagrams...

Sexual Reproduction and Variation

You saw on page 36 that everyone is <u>slightly different</u>, partly because of their different <u>environments</u> but also partly because everyone has different <u>genes</u> (apart from identical twins). So, how come we all have different genes? Well, it's partly because of how <u>sexual reproduction</u> works, and partly due to <u>mutation</u>.

Sexual Reproduction Produces Genetically Different Cells

1) <u>Sexual reproduction</u> is where genetic information from <u>two</u> organisms (a <u>father</u> and a <u>mother</u>) is combined to produce offspring which are <u>genetically different</u> to either parent.

2) In sexual reproduction the mother and father produce <u>gametes</u> — e.g. <u>egg</u> and <u>sperm</u> cells in animals.

3) In humans, each gamete contains <u>23 chromosomes</u> — <u>half</u> the number of chromosomes in a normal cell. (Instead of having <u>two</u> of each chromosome, a <u>gamete</u> has just <u>one</u> of each.)

4) The <u>egg</u> (from the mother) and the <u>sperm</u> cell (from the father) then <u>fuse together</u> (fertilisation) to form a cell with the <u>full number</u> of chromosomes (<u>half from the father</u>, <u>half from the mother</u>).

> <u>SEXUAL REPRODUCTION</u> involves the fusion of male and female gametes.
> Because there are <u>TWO</u> parents, the offspring contains <u>a mixture of their parents' genes</u>.

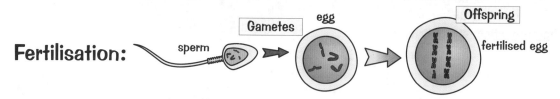

Fertilisation: sperm Gametes egg Offspring fertilised egg

5) This is why the offspring <u>inherits features</u> from <u>both parents</u> — it's received a mixture of chromosomes from its mum and its dad (and it's the chromosomes that decide how you turn out).

6) This is why <u>sexual</u> reproduction produces more variation than <u>asexual</u> reproduction (see page 41).

Mutations Are Changes to the Genetic Code

1) Occasionally a gene may <u>mutate</u>. Mutations <u>change the sequence</u> of the <u>DNA bases</u> (see page 53) This could <u>stop the production</u> of a <u>protein</u>, or it might mean a <u>different</u> protein is produced instead. This can lead to <u>new characteristics</u>, <u>increasing variation</u>.

2) Mutations can happen <u>spontaneously</u> — when a chromosome doesn't quite copy itself properly. However, the chance of mutation is <u>increased</u> by exposing yourself to:
- nuclear radiation, X-rays or ultraviolet light,
- <u>chemicals</u> called <u>mutagens</u>. (<u>Cigarette smoke</u> contains <u>mutagens</u>.)
  If the mutations lead to cancer then the chemicals are called carcinogens.

No no! not me!

3) Mutations are usually harmful.
- If a mutation occurs in <u>reproductive cells</u>, the offspring might develop <u>abnormally</u> or <u>die</u>.
- If a mutation occurs in body cells, the mutant cells may start to <u>multiply</u> in an <u>uncontrolled</u> way and <u>invade</u> other parts of the body (which is <u>cancer</u>).

4) <u>Very occasionally</u>, mutations are beneficial and give an organism a survival <u>advantage</u>, so it can live on in conditions where the others die. This is <u>natural selection</u> at work.
For example, a mutation in a bacterium might make it <u>resistant to antibiotics</u>. If this mutant gene is passed on, you might get a <u>resistant "strain"</u> of bacteria, which antibiotics can't kill.

You need to reproduce these facts in the exam...

The main messages on this page are that: 1) <u>Sexual</u> reproduction needs <u>two</u> parents to form offspring that are <u>genetically different</u> to the parents, so there's <u>lots</u> of variation. 2) <u>Mutations</u> can also lead to an increase in variation and can occasionally be <u>beneficial</u> — although they're more likely to be <u>harmful</u>.

Genetic Diagrams

In the exam they could ask you about the inheritance of <u>any characteristic</u> controlled by a single gene. Luckily the <u>principle</u> is the same no matter what the characteristic...

Alleles Are Different Versions of the Same Gene

1) Most of the time you have <u>two</u> of each gene (i.e. two alleles) — one from each parent.

2) If the alleles are different you have instructions for two different versions of a characteristic (e.g. blue eyes or brown eyes), but you only show one version of the two (e.g. brown eyes). The version of the characteristic that appears is caused by the <u>dominant allele</u>. The other allele is said to be <u>recessive</u>.

3) In genetic diagrams <u>letters</u> are used to represent <u>genes</u>. <u>Dominant</u> alleles are always shown with a <u>capital letter</u> (e.g. 'C') and <u>recessive</u> alleles with a <u>small letter</u> (e.g. 'c').

4) If you're <u>homozygous</u> for a trait you have <u>two alleles the same</u> for that particular gene, e.g. CC or cc. If you're <u>heterozygous</u> for a trait you have <u>two different alleles</u> for that particular gene, e.g. Cc.

You Need to be Able to Construct and Explain Genetic Diagrams

Imagine you're cross-breeding <u>hamsters</u>, and that some have a normal, boring disposition while others have a leaning towards crazy acrobatics. And suppose you know that the behaviour is due to one gene...

Let's say that the allele which causes the crazy nature is <u>recessive</u> — so use a '<u>b</u>'.
And normal (boring) behaviour is due to a <u>dominant allele</u> — call it '<u>B</u>'.

1) For an organism to display a <u>recessive</u> characteristic, <u>both</u> its alleles must be <u>recessive</u> — so a crazy hamster must have the alleles 'bb' (i.e. it must be homozygous for this trait).

2) However, a <u>normal hamster</u> could be BB (homozygous) or Bb (heterozygous), because the dominant allele (B) <u>overrules</u> the recessive one (b).

So if you cross a <u>thoroughbred crazy hamster</u>, genetic type bb, with a <u>thoroughbred normal hamster</u>, BB, you get this...

Parents: normal and boring parent   crazy parent

Parents' alleles: (BB)   (bb)

Gametes' alleles: (B)   (B)   (b)   (b)

Possible combinations of alleles in offspring: (Bb)   (Bb)   (Bb)   (Bb)

They're <u>all</u> normal and boring.

The lines show <u>all</u> the <u>possible</u> ways the parents' alleles <u>could</u> combine.

Remember, only <u>one</u> of these possibilities would <u>actually happen</u> for any one offspring.

When you breed two organisms together to look at one characteristic it's called a <u>MONOHYBRID CROSS</u>.

If two of these <u>offspring</u> now breed they will produce a <u>new combination</u> of kids.

This time, there's a <u>75%</u> chance of having a normal, boring hamster, and a <u>25%</u> chance of a crazy one.

(To put that another way... you'd expect a 3:1 ratio of normal:crazy hamsters.)

Parents: normal and boring   normal and boring

Parents' alleles: (Bb)   (Bb)

Gametes' alleles: (B)   (b)   (B)   (b)

Possible combinations of alleles in offspring: (BB)   (Bb)   (Bb)   (bb)
normal   normal   normal   <u>crazy!</u>

I think my sister has the wild and scratty allele too...

Interestingly (maybe), you can draw a similar diagram to show the probability of having a <u>boy or a girl</u>. (It's not a single gene that determines sex, but a <u>chromosome</u> — the diagram's exactly the same though.) Women have two <u>X-chromosomes</u>, whereas men have <u>an X and a Y</u>, the Y-chromosome being <u>dominant</u>. It turns out the odds are <u>50:50</u>. Have a go at drawing the genetic diagram to show this.

Genetic Disorders

Sometimes an allele might be <u>faulty</u> and <u>not work properly</u>. It can cause more than a few problems...

Genetic Disorders are Caused by Faulty Alleles

A faulty allele <u>could</u> have either of these effects...

1) The faulty gene could <u>directly</u> cause a <u>genetic disorder</u>. Cystic fibrosis, haemophilia, red-green colour blindness (and many other disorders) are all caused by faulty genes.

2) The faulty gene may not cause a problem in itself, but it could mean a <u>predisposition</u> to certain health problems — meaning it's <u>more likely</u> (though not definite) that you'll suffer from them in the <u>future</u> (e.g. some genes predispose people to getting <u>breast cancer</u>).

Cystic Fibrosis is Caused by a Recessive Allele

<u>Cystic fibrosis</u> is a <u>genetic disorder</u> of the <u>cell membranes</u>. It results in the body producing a lot of thick sticky <u>mucus</u> in the <u>air passages</u> and in the <u>pancreas</u>.

1) The allele which causes cystic fibrosis is a <u>recessive allele</u>, 'f', carried by about <u>1 person in 25</u>.

2) Because it's recessive, people with only <u>one copy</u> of the allele <u>won't</u> have the disorder — they're known as <u>carriers</u>.

3) For a child to have a chance of inheriting the disorder, <u>both parents</u> must be either <u>carriers</u> or <u>sufferers</u>.

4) As the diagram shows, there's a <u>1 in 4 chance</u> of a child having the disorder if <u>both</u> parents are <u>carriers</u>.

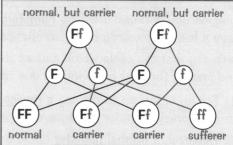

Knowing About Genetic Disorders Opens Up a Whole Can of Worms

Knowing there are inherited conditions in your family raises <u>difficult issues</u>:

- Should all family members be <u>tested</u> to see if they're carriers? Some people might prefer <u>not to know</u>, but is this <u>fair</u> on any partners or future children they might have?

- Is it <u>right</u> for someone who's at risk of passing on a genetic condition to have <u>children</u>? Is it <u>fair</u> to put them under pressure <u>not to</u>, if they decide they want children?

- It's possible to <u>test</u> a foetus for some genetic conditions while it's still in the <u>womb</u>. But if the test is positive, is it right to <u>terminate</u> the pregnancy? The family might not be able to <u>cope</u> with a sick or disabled child, but why should that child have a lesser <u>right to life</u> than a healthy child? Some people think abortion is <u>always wrong</u> under any circumstances.

Gene Therapy is being Developed to Treat Genetic Disorders

Gene therapy means <u>correcting faulty genes</u> — usually a <u>healthy copy</u> of the gene is added.

Example — Treating Cystic Fibrosis

1) At the moment scientists are trying to cure cystic fibrosis (CF) with gene therapy. One method being trialled is the use of a virus to insert a <u>healthy copy</u> of the gene into cells in the airways.

2) There are still problems — for example, at the moment the effect wears off after a <u>few days</u>. But there are big hopes that gene therapy will one day mean CF can be treated effectively.

3) However, since this kind of gene therapy involves only body cells (and not reproductive cells), the faulty gene would still be passed on to children.

Unintentional mooning — caused by faulty jeans...

On a related note... the <u>Human Genome Project</u> aimed to map all the genes in a human. This has now been completed, and the results could help with future gene therapies. Exciting stuff.

Cloning

Sexual reproduction was covered on page 38 — but that's not the only way of doing things...

Asexual Reproduction Produces Genetically Identical Cells

1) An ordinary cell can make a new cell by simply dividing in two. The new cell has exactly the same genetic information (i.e. genes) as the parent cell — this is known as asexual reproduction.

> In ASEXUAL REPRODUCTION there is only ONE parent, and the offspring has identical genes to the parent (i.e. there's no variation between parent and offspring, so they're clones — see also p42).

2) Here's how it works...

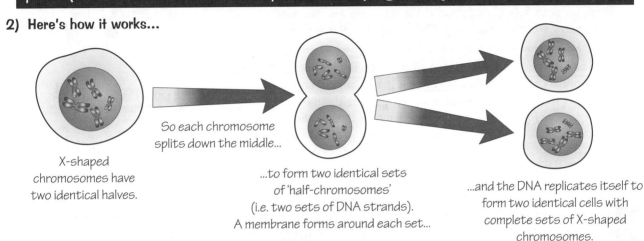

So each chromosome splits down the middle...

X-shaped chromosomes have two identical halves.

...to form two identical sets of 'half-chromosomes' (i.e. two sets of DNA strands). A membrane forms around each set...

...and the DNA replicates itself to form two identical cells with complete sets of X-shaped chromosomes.

3) This is how all plants and animals grow and produce replacement cells.

4) Some organisms also produce offspring using asexual reproduction, e.g. bacteria and certain plants.

Plants Can Be Cloned from Cuttings and by Tissue Culture

Asexual reproduction can be used to clone plants. And it's not all high-tech crazy science stuff either — gardeners have been using cuttings since before your gran was knee-high to a grasshopper.

CUTTINGS

1) Gardeners can take cuttings from good parent plants, and then plant them to produce genetically identical copies (clones) of the parent plant.

2) These plants can be produced quickly and cheaply.

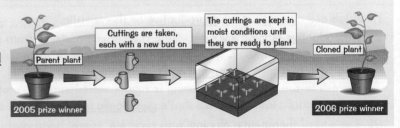

Parent plant — 2005 prize winner

Cuttings are taken, each with a new bud on

The cuttings are kept in moist conditions until they are ready to plant

Cloned plant — 2006 prize winner

TISSUE CULTURE

This is where a few plant cells are put in a growth medium with hormones, and they then grow into new plants — clones of the parent plant. The advantages of using tissue culture are that you can make new plants very quickly, in very little space, and you can grow all year.

The disadvantage to both these methods is a 'reduced gene pool' (see next page).

Reproduction doesn't have to involve sex — ask any spider plant...

You do need to know all the details on this page, I'm afraid. Certain plants can reproduce asexually, e.g. potatoes, strawberries, daffodils and chlorophytum (spider plants). Spider plants grow tufty bits at the end of their shoots called plantlets — each plantlet is a clone of the original plant. The clone grows roots and becomes a new plant.

Plantlets

Cloning

If you've cloned a cow before you won't need to learn this page, if not you'd better read on.

You Need to Know About Embryo Transplants in Cows

Normally, farmers only breed from their best cows and bulls. However, traditional methods would only allow the prize cow to produce one new offspring each year. These days the whole process has been transformed using embryo transplants:

1) Sperm cells are taken from the prize bull. They can also be frozen and used at a later date. Prize cows are given hormones to make them produce lots of eggs and these egg cells are removed. The sperm are then used to artificially fertilise the egg cells.

2) Now, here's the clever bit. Each fertilised egg starts dividing to give a new calf, but in the early stages all the cells are identical. These balls of identical cells are called embryos. The scientists check the embryos for their sex, screen them for genetic defects and then split them to give lots of individual cells. Each of these cells carries on growing to give a new embryo which is a clone of the original one.

"Nurse — the screens!"

3) The offspring are clones of each other, not clones of their parents.

4) These embryos are implanted into other cows, called 'surrogate mothers', to grow. Or they can be frozen and used later.

Advantages

a) Hundreds of "ideal" offspring can be produced every year from the best bull and cow.

b) The original prize cow can keep producing prize eggs all year round.

Disadvantages

The main problem is that the same alleles keep appearing (and many others are lost). So there's a greater risk of genetic disorders, and a disease could wipe out an entire population if there are no resistant alleles (see page 39).

Oh Eck!

Adult Cell Cloning is Another Way to Make a Clone...

1) Adult cell cloning can be reproductive or therapeutic.

2) Reproductive cloning involves taking the genetic material from an adult cell to make a new organism that is a clone of that adult. It's now been done in lots of mammals, including sheep, horses and cats. Basically, you take an egg cell and remove its genetic material. A complete set of chromosomes from the cell of the adult you're cloning is then inserted into the 'empty' egg cell, which grows into an embryo and eventually into an animal that's genetically identical to the original adult.

3) The aim of therapeutic cloning is to produce 'spare' body parts for disease sufferers without them being rejected by the sufferer's immune system. A cloned embryo is created that is genetically identical to the sufferer and special cells (embryonic stem cells) that can become any cell in the body are extracted from it.

4) Some people think it's unethical to do this because the embryos used to provide the stem cells are destroyed. Fusion cloning could avoid this. Here, an adult cell is joined to an already existing (but genetically different) embryonic stem cell. The result has the properties of a stem cell but the same genes as the adult.

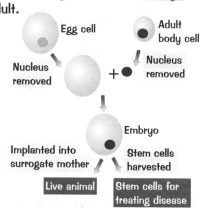

Egg cell — Adult body cell

Nucleus removed + Nucleus removed

Embryo

Implanted into surrogate mother — Stem cells harvested

Live animal | Stem cells for treating disease

See page 76 for more on adult cell cloning.

Embryo transplants — I'm glad they didn't do that with my sister...

So if you're cloning an adult cell, you can have reproductive cloning, where the aim is to make a brand new organism, or therapeutic cloning, where you don't make a new organism, just harvest the stem cells.

Genetic Engineering

Scientists can now <u>add</u>, <u>remove</u> or <u>change</u> an organism's <u>genes</u> to alter its characteristics.

Genetic Engineering Uses Enzymes to Cut and Paste Genes

The basic idea is to move <u>useful genes</u> from one organism's chromosomes into the cells of another...

1) A useful gene is "<u>cut</u>" from one organism's chromosome using <u>enzymes</u>.

2) <u>Enzymes</u> are then used to <u>cut</u> another organism's chromosome and to <u>insert</u> the useful gene. This technique is called <u>gene splicing</u>.

3) Scientists use this method to do all sorts of things — for example, the human insulin gene can be inserted into <u>bacteria</u> to <u>produce human insulin</u>:

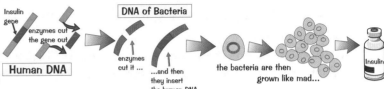

Insulin gene — enzymes cut the gene out — Human DNA — enzymes cut it ... — ...and then they insert the human DNA — DNA of Bacteria — the bacteria are then grown like mad... — Insulin

Genes can be Transferred into Animals and Plants

The same method can be used to <u>transfer useful genes</u> into <u>animals</u> and <u>plants</u> at the <u>very early stages</u> of their development (i.e. shortly after <u>fertilisation</u>). This has (or could have) some really <u>useful applications</u>.

1) <u>Genetically modified (GM) plants</u> have been developed that are <u>resistant to viruses</u> and <u>herbicides</u> (chemicals used to kill weeds). And <u>long-life</u> tomatoes can be made by changing the gene that causes the fruit to ripen.

2) Genes can also be inserted into <u>animal embryos</u> so that the animal grows up to have more <u>useful characteristics</u>. For example, <u>sheep</u> have been genetically engineered to produce substances (e.g. drugs) in their <u>milk</u> that can be used to treat <u>human diseases</u>.

3) <u>Genetic disorders</u> like cystic fibrosis are caused by faulty genes. Scientists are trying to cure these disorders by <u>inserting working genes</u> into sufferers. This is called <u>gene therapy</u> — see page 40.

But Genetic Engineering is a Controversial Topic...

So, genetic engineering is an <u>exciting new area in science</u> which has the <u>potential</u> to solve many of our problems (e.g. treating diseases, more efficient food production etc.) but not everyone likes it.

1) Some people <u>strongly believe</u> that we shouldn't go <u>tinkering about with genes</u> because it's <u>not natural</u>.

2) There are <u>worries</u> that changing an organism's genes might <u>accidentally</u> create unforeseen <u>problems</u> — which could then get passed on to <u>future generations</u>.

There Are Pros and Cons with GM Crops

1) Some people say that growing GM crops will affect the number of <u>weeds</u> and <u>flowers</u> (and therefore <u>wildlife</u>) that usually lives in and around the crops — <u>reducing</u> farmland <u>biodiversity</u>.

2) Not everyone is convinced that GM crops are <u>safe</u>. People are worried they may develop <u>allergies</u> to the food — although there's probably no more risk for this than for eating normal foods.

3) A big concern is that <u>transplanted genes</u> may get out into the <u>natural environment</u>. For example, the <u>herbicide resistance</u> gene may be picked up by weeds, creating a new '<u>superweed</u>' variety.

4) On the plus side, GM crops can <u>increase the yield</u> of a crop, making more food.

5) People living in developing nations often lack <u>nutrients</u> in their diets. GM crops could be <u>engineered</u> to contain nutrients that are <u>missing</u>. For example, they're testing 'golden rice' that contains beta-carotene — lack of this substance can cause <u>blindness</u>.

I say it's great.

6) GM crops are already being used elsewhere in the world (not the UK), often <u>without any problems</u>.

If only there was a gene to make revision easier...

It's up to the <u>Government</u> to weigh up all the <u>evidence</u> before <u>making a decision</u> on how this knowledge is used. All scientists can do is make sure the Government has all the information it needs.

Revision Summary for Section Four

There's a lot to remember in this section and quite a few of the topics are controversial, e.g. cloning, genetic engineering, and so on. You need to know all sides of the story, as well as all the facts... so, here are some questions to help you. If you get any wrong, go back and learn that bit again.

1) What are the two types of variation?

2) Describe the relative importance of each type for plants and for animals.

3) List four features of animals which aren't affected at all by their environment, and three which are.

4) Draw a set of diagrams showing the relationship between: cell, nucleus, chromosomes, DNA.

5) What is an allele?

6) How many pairs of chromosomes does a normal human cell nucleus contain?
   Which cells, found in every adult human, have a different number in their nucleus?

7) Explain why sexual reproduction results in offspring that are genetically different from either parent.

8) Name three things that cause genetic mutations.

9) Give an example of how a genetic mutation could be:

   a) harmful,   b) beneficial.

10) What does it mean if you are homozygous for a particular trait?

11)* Draw a genetic diagram for a cross between a man who has blue eyes (bb) and
    a woman who has green eyes (Bb). The gene for blue eyes (b) is recessive.

12) Describe two ways in which a faulty gene could lead to health problems.

13) It is now possible to test whether or not you are a carrier for many genetic disorders.
    Outline some of the reasons in favour of this testing, and also suggest reasons why an individual
    might prefer not to be tested.

14) What's the basic idea behind gene therapy?

15) Give a definition of asexual reproduction.

16) Describe how to make plant clones from:
    a) cuttings,   b) tissue culture.

17) Give an advantage and a disadvantage of producing cloned plants.

18) Describe two different ways to clone an animal.

19) What are the advantages and disadvantages of cloning using embryo transplants?

20) Give an account of the important stages of genetic engineering.

21) State two examples of useful applications of genetic engineering.

22) Why are some people concerned about genetic engineering?

* Answer on page 140

There's Too Many People

We have an <u>impact</u> on the world around us — and the <u>more humans</u> there are, the bigger the impact.

There are <u>Six Billion People</u> in the World...

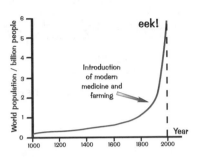

1) The <u>population</u> of the world is currently <u>rising</u> very quickly, and it's not slowing down — look at the graph...

2) This is mostly due to modern <u>medicine</u> and <u>farming</u> methods, which have <u>reduced</u> the number of <u>people dying</u> from <u>disease</u> and <u>hunger</u>.

3) This is great for all of us <u>humans</u>, but it means we're having a <u>bigger effect</u> on the <u>environment</u> we live in...

...<u>With Increasing Demands</u> on the <u>Environment</u>

When the <u>Earth's population</u> was much smaller, the effects of <u>human activity</u> were usually <u>small</u> and <u>local</u>. Nowadays though, our actions can have a far more <u>widespread</u> effect.

1) Our rapidly increasing <u>population</u> puts pressure on the <u>environment</u>, as we take the resources we need to <u>survive</u>.

2) But people around the world are also demanding a higher <u>standard of living</u> (and so demand luxuries to make life more comfortable — cars, computers, etc.). So we use more <u>raw materials</u> (e.g. oil to make plastics), but we also use more <u>energy</u> for the manufacturing processes. This all means we're taking more and more <u>resources</u> from the environment more and more <u>quickly</u>.

3) Unfortunately, many raw materials are being used up more quickly than they're being replaced. So if we carry on like we are, one day we're going to <u>run out</u>.

We're Also Producing More Waste

As we make more and more things we produce more and more <u>waste</u>. And unless this waste is properly handled, more <u>harmful pollution</u> will be caused. This affects water, land and air.

Water | <u>Sewage</u> and <u>toxic chemicals</u> from industry can pollute lakes, rivers and oceans, affecting the plants and animals that rely on them for survival (including humans). And the chemicals used on land (e.g. fertilisers) can be washed into water.

Land | We use <u>toxic chemicals</u> for farming (e.g. pesticides and herbicides). We also bury <u>nuclear waste</u> underground, and we dump a lot of <u>household waste</u> in landfill sites.

Air | <u>Smoke</u> and <u>gases</u> released into the atmosphere can pollute the air (see page 48 for more info). For example, <u>sulfur dioxide</u> can cause <u>acid rain</u>.

More People Means Less Land for Plants and Other Animals

Humans also <u>reduce</u> the amount of <u>land and resources</u> available to other <u>animals</u> and <u>plants</u>. The <u>four main human activities</u> that do this are:

1) <u>Building</u>

2) <u>Farming</u>

3) <u>Dumping Waste</u>

4) <u>Quarrying</u>

More people, more mess, less space, less resources...

Well, I feel guilty. I don't know about you. Not only are we taking more and more land for building and farming and quarrying but we're also polluting what little there is left over. In the exam you might be given some <u>data</u> about <u>environmental impact</u>, so make sure you understand what's going on...

The Greenhouse Effect

The greenhouse effect is always in the news. We need it, since it makes Earth a suitable temperature for living on. But unfortunately it's starting to trap more heat than is necessary.

Carbon Dioxide and Methane Trap Heat from the Sun

1) The temperature of the Earth is a balance between the heat it gets from the Sun and the heat it radiates back out into space.

2) Gases in the atmosphere absorb most of the heat that would normally be radiated out into space, and re-radiate it in all directions (including back towards the Earth).

3) If this didn't happen, then at night there'd be nothing to keep any heat in, and we'd quickly get very cold indeed.

4) There are several different gases in the atmosphere which help keep the heat in. They're called "greenhouse gases" (oddly enough) and they include carbon dioxide and methane.

5) Humans release carbon dioxide into the atmosphere as part of our everyday lives — e.g. as we burn fossil fuels in power stations or cars.

6) The Earth is gradually heating up because of the increasing levels of greenhouse gases — this is global warming. Global warming is a type of climate change and causes other types of climate change, e.g. changing rainfall patterns.

This is what happens in a greenhouse. The sun shines in, and the glass helps keeps some of the heat in.

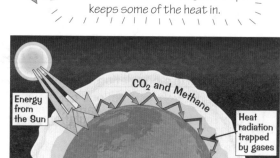

Human Activity Produces Lots of Carbon Dioxide

1) Humans release carbon dioxide into the atmosphere all the time as part of our everyday lives — in car exhausts, industrial processes, as we burn fossil fuels etc.

2) People around the world are also cutting down large areas of forest (deforestation) for timber and to clear land for farming — and this activity affects the level of carbon dioxide in the atmosphere in various ways:

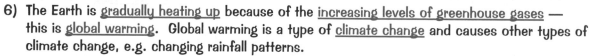

- Carbon dioxide is released when trees are burnt to clear land. (Carbon in wood is 'locked up' and doesn't contribute to atmospheric pollution — until it's released by burning.)
- Microorganisms feeding on bits of dead wood release CO_2 as a waste product of respiration.
- Cutting down loads of trees means that the amount of carbon dioxide removed from the atmosphere during photosynthesis is reduced.

So we're putting more CO_2 into the atmosphere and taking less out.

Methane is Also a Problem...

1) Methane gas is also contributing to the greenhouse effect.

2) It's produced naturally from various sources, e.g. rotting plants in marshland.

3) However, two 'man-made' sources of methane are on the increase:

    a) Rice growing

    b) Cattle rearing — it's the cows' "pumping" that's the problem, believe it or not.

Methane is a stinky problem but an important one...

Global warming is rarely out of the news. Scientists accept that it's happening and that human activity has caused most of the recent warming. However, they don't know exactly what the effects will be...

Climate Change

The Earth is getting warmer. Climate scientists are now trying to work out what the effects of global warming might be — sadly, it's not as simple as everyone having nicer summers.

The Consequences of Global Warming Could be Pretty Serious

There are several reasons to be worried about global warming. Here are a few:

1) As the sea gets warmer, it expands, causing sea level to rise. Sea level has risen a little bit over the last 100 years. If it keeps rising it'll be bad news for people living in low-lying places like the Netherlands, East Anglia and the Maldives — they'd be flooded.

2) Higher temperatures make ice melt. Water that's currently 'trapped' on land (as ice) runs into the sea, causing sea level to rise even more.

3) Global warming has changed weather patterns in many parts of the world. It's thought that many regions will suffer more extreme weather because of this, e.g. longer, hotter droughts. Hurricanes form over water that's warmer than 27 °C — so with more warm water, you'd expect more hurricanes.

4) Changing weather patterns also affect food production — some regions are now too dry to grow food, some too wet. This will get worse as temperature increases and weather patterns change more.

5) The climate is a very complicated system. For instance, if the ice melts, there's less white stuff around to reflect the sun's rays out to space, so maybe we'll absorb more heat and get even warmer. But... when the sea's warmer, more water evaporates, making more clouds — and they reflect the Sun's rays, so maybe we'd cool down again. So it's hard to predict exactly what will happen, but lots of people are working on it, and it's not looking too good.

You Need to Weigh the Evidence Before Making Judgements

1) To find out how our climate is changing, scientists are busy collecting data about the environment.

2) For instance, we're using satellites to monitor snow and ice cover, and to measure the temperature of the sea surface. We're recording the temperature and speed of the ocean currents, to try and detect any changes. Automatic weather stations are constantly recording atmospheric temperatures.

3) All this data is only useful if it covers a wide enough area and a long enough time scale.

4) Generally, observations of a very small area aren't much use. Noticing that your local glacier seems to be melting does not mean that ice everywhere is melting, and it's certainly not a valid way to show that global temperature is changing. (That would be like going to Wales, seeing a stripy cow and concluding that all the cows in Wales are turning into zebras.) Looking at the area of ice cover over a whole continent, like Antarctica, would be better.

5) The same thing goes for time. It's no good going to the Arctic, seeing four polar bears one week but only two the next week and concluding that polar bears are dying out because the ice is disappearing. You need to do your observations again and again, year after year.

6) Scientists can make mistakes — so don't take one person's word for something, even if they've got a PhD. But if lots of scientists get the same result using different methods, it's probably right. That's why most governments around the world are starting to take climate change seriously.

Climate control — it's optional on most luxury cars...

We humans have created some big environmental problems for ourselves. Many people, and some governments, think we ought to start cleaning up the mess. Scientists can help, mainly in understanding the problems and suggesting solutions, but it's society as a whole that has to do something.

Air Pollution

Other polluting gases released through human activity can cause havoc in the air.

CFCs Cause Ozone Depletion

1) CFCs (chlorofluorocarbons) used to be used in aerosols, fridges, air-conditioning units, and polystyrene foam.
2) They break down ozone in the upper atmosphere.
3) This allows more harmful UV rays to reach the Earth's surface.
4) Being exposed to more UV rays will increase the risk of skin cancer (although this can be reduced with suncream). Australia has high levels of skin cancer because it is under an ozone hole.
5) The increase in UV rays might also kill plankton in the sea — this could have a massive effect on the sea ecosystem because plankton are at the bottom of the food chain. Scientists predict that fish levels will drop (meaning, among other things, less food for us to eat).

Carbon Monoxide is Poisonous

1) When fossil fuels are burnt without enough air supply they produce the gas carbon monoxide (CO).
2) It's a poisonous gas. If it combines with red blood cells, it prevents them from carrying oxygen.
3) Carbon monoxide's mostly released in car emissions. Most modern cars are fitted with catalytic converters that oxidise the carbon monoxide (to make carbon dioxide), decreasing the amount that's released into the atmosphere.

Acid Rain is Caused by Sulfur Dioxide and Oxides of Nitrogen

1) As well as releasing CO_2, burning fossil fuels releases other harmful gases. These include sulfur dioxide and various nitrogen oxides.
2) The sulfur dioxide (SO_2) comes from sulfur impurities in the fossil fuels. The nitrogen oxides are made in a reaction between nitrogen and oxygen in the air, caused by the heat of the burning.
3) When these gases mix with rain clouds they form dilute sulfuric acid and dilute nitric acid.
4) This then falls as acid rain.
5) Internal combustion engines in cars and power stations are the main causes of acid rain.

Acid Rain Kills Fish and Trees and Damages Statues

1) Acid rain can cause a lake to become more acidic. This has a severe effect on the lake's ecosystem. Many organisms are sensitive to changes in pH and can't survive in more acidic conditions. Many plants and animals die.
2) Acid rain can kill trees.
3) Acid rain damages limestone buildings and statues.

It's raining, it's pouring — quick, cover the rhododendron...

Exam questions on this topic may give you a graph or table to interpret. If they ask you to describe the data, just say what you see (e.g. the number of species in a lake decreases with increased acidity of the water). But if they ask you to explain the data, you're going to need some scientific knowledge to suggest why the data shows that trend (e.g. some species can't survive in acidic conditions).

Sustainable Development

There is a growing feeling among scientists and politicians that if we carry on behaving as we are, we may end up causing huge problems for future generations...

Sustainable Development Needs Careful Planning

1) Human activities can damage the environment (e.g. pollution). And some of the damage we do can't easily be repaired (e.g. the destruction of the rainforests).

2) We're also placing greater pressure on our planet's limited resources (e.g. oil is a non-renewable resource so it will eventually run out).

3) This means that we need to plan carefully to make sure that our activities today don't mess things up for future generations — this is the idea behind sustainable development...

> **SUSTAINABLE DEVELOPMENT** meets the needs of today's population without harming the ability of future generations to meet their own needs.

4) This isn't easy — it needs detailed thought at every level to make it happen. For example, governments around the world will need to make careful plans. But so will the people in charge at a regional level.

Reduction in Biodiversity Could Be a Big Problem

Biodiversity is the variety of different species in an area — the more species, the higher the biodiversity. Ecosystems (especially tropical rainforests) can contain a huge number of different species, so when a habitat is destroyed there is a danger of many species becoming extinct — biodiversity is reduced. This causes a number of lost opportunities for humans and problems for those species that are left:

1) There are probably loads of useful products that we will never know about because the organisms that produced them have become extinct. Newly discovered plants and animals are a great source of new foods, new fibres for clothing and new medicines, e.g. the rosy periwinkle flower from Madagascar has helped treat Hodgkin's disease (a type of cancer), and a chemical in the saliva of a leech has been used to help prevent blood clots during surgery.

2) Loss of one or more species from an ecosystem unbalances it, e.g. the extinct animal's predators may die out or be reduced. Loss of biodiversity can have a 'snowball effect' which prevents the ecosystem providing things we need, such as rich soil, clean water, and the oxygen we breathe.

Human Impact can be Measured Using Indicator Species

Getting an accurate picture of the human impact on the environment is hard. But one technique that's used involves indicator species.

1) Some organisms are very sensitive to changes in their environment and so can be studied to see the effect of human activities — these organisms are known as indicator species.

2) For example, air pollution can be monitored by looking at particular types of lichen, which are very sensitive to levels of sulfur dioxide in the atmosphere (and so can give a good idea about the level of pollution from car exhausts, power stations, etc.). The number and type of lichen at a particular location will indicate how clean the air is (e.g. the air is clean if there are lots of lichen).

3) If raw sewage is released into a river, the bacterial population in the water increases and uses up the oxygen. Animals like mayfly larvae are good indicators for water pollution, because they are very sensitive to the level of oxygen in the water. If you find mayfly larvae in a river, it indicates that the water is clean.

Teenagers are an indicator species — not found in clean rooms...

In the exam, make sure you remember the details about the environmental problems that development can cause. If you get an essay-type question, stick 'em in to show off your 'scientific knowledge'.

Conservation and Recycling

Conservation and recycling are all about what humans can do to <u>reduce our impact</u> on the environment.

Conservation is Important for Protecting Food, Nature and Culture

1) Conservation measures <u>protect species</u> by <u>maintaining</u> their <u>habitats</u> and <u>protecting</u> them from <u>poachers</u> and <u>over-hunting</u> / <u>over-harvesting</u>.

2) There are several <u>reasons</u> why it's important to <u>conserve species and natural habitats</u>:

- <u>PROTECTING ENDANGERED SPECIES</u> — Many species are now <u>endangered</u>, often due to hunting and the <u>destruction</u> of their <u>habitats</u>. They need to be protected to stop them becoming extinct.
- <u>PROTECTING THE HUMAN FOOD SUPPLY</u> — Overfishing has greatly <u>reduced fish stocks</u> in the <u>sea</u>. Conservation measures (e.g. <u>quotas</u> on how many fish can be caught) encourage the survival and <u>growth</u> of fish stocks. This <u>protects the food supply</u> for future generations.
- <u>MAINTAINING BIODIVERSITY</u> — See previous page.

Example: Woodland Conservation

Conservation measures in a woodland habitat may include:

1) Coppicing — This is an ancient form of woodland management. It involves <u>cutting trees</u> down to just above ground level. The <u>stumps</u> sprout <u>straight, new stems</u> which can be regularly harvested.

2) Reforestation — Where forests have been cut down in the past, they can be <u>replanted</u> to <u>recreate</u> the <u>habitat</u> that has been lost.

3) Replacement planting — This is when <u>new trees</u> are <u>planted</u> at the <u>same rate</u> that others are <u>cut down</u>. So the total number of trees remains the same.

Recycling Conserves Our Natural Resources

If materials aren't recycled they get <u>thrown away as waste</u>. This means that:

1) There is <u>more waste</u>, so <u>more land</u> has to be used for <u>landfill sites</u> (waste dumps). Some waste is <u>toxic</u> (poisonous), so this also means more polluted land.

2) <u>More materials</u> have to be <u>manufactured</u> or <u>extracted</u> to make new products (rather than recycling existing ones) — using up more of the Earth's resources and more energy.

Recycling uses up less of the Earth's <u>natural resources</u>. <u>Recycling processes</u> usually use <u>less energy</u> and create <u>less pollution</u> than manufacturing or extracting materials from scratch. Recyclable materials include metals, paper, plastics and glass.

There are Some Problems with Recycling

1) Recycling still <u>uses energy</u>, e.g. for <u>collecting</u>, <u>sorting</u>, <u>cleaning</u> and <u>processing waste</u>.

2) Some waste materials can be difficult and <u>time-consuming to sort</u> out, e.g. different types of <u>plastic</u> have to be separated from each other before they can be recycled.

3) The <u>equipment needed</u> for recycling can be <u>expensive</u>, e.g. equipment for sorting plastics automatically.

4) In some cases, the <u>quality</u> of recycled materials <u>isn't as good</u> as new materials, e.g. recycled paper.

5) <u>Some materials</u> can only be <u>recycled</u> a <u>limited number of times</u> (e.g. plastics, paper). Others can be recycled indefinitely though (e.g. aluminium).

Recycling — do the Tour de France twice...

The organisms in a habitat are dependent on each other, e.g. for food. You need to <u>protect all species</u> — animals, trees, fungi, bacteria... because if one of them dies out it affects the others (see p49).

Revision Summary for Section Five

So, no cheerful stuff here. It's all doom and gloom. All these humans are hogging the land, and filling the air with evil pollutants. In fact, planet Earth would be much better off if we all emigrated to Mars.

Anyway, it's time for you to answer some questions on the woes of the world. You should know the drill by now. Try all the questions and if you get any wrong it's straight back to the page on that topic — do not pass go, do not collect £200. Before long, you'll be an environmental guru and won't be fazed by any question that turns up in the exam.

1) What is happening to the world's population? What is largely responsible for this trend?

2) Suggest three ways in which a rising population is affecting the environment.

3) What are the main four human activities that use up land?

4) Name two important greenhouse gases. Why are they called 'greenhouse' gases?

5) Draw and label a diagram to explain the greenhouse effect.

6) Give three ways that deforestation adds to the greenhouse effect.

7) Give two human activities (apart from deforestation) that release carbon dioxide.

8) Name two human activities that are increasing the release of methane.

9) What problems could global warming cause?

10) Give three different types of data that scientists are collecting to try and determine how the climate is changing as a result of global warming.

11)* Read the statement below and consider how valid it is.

> The Malaspina Glacier in Alaska is losing over 2.7 km³ of water each year. This proves global warming is happening.

12) Explain how CFCs could end up having a serious impact on the Earth's oceans.

13) Which gases cause acid rain? How are they produced?

14) Describe the damage that can be caused by acid rain.

15) Define sustainable development.

16)* The graph on the right shows human population growth and an estimate of the number of species that have become extinct between 1800 and 2000.

   a) How are the size of the human population and the number of extinct species related?

   b) Suggest a reason for this relationship.

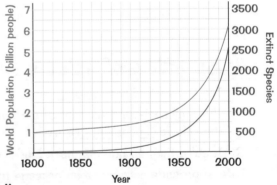

17) Explain how lichen can be used as an indicator of air pollution.

18) What does it suggest about the cleanliness of the water if you find mayfly larvae in a river?

19) Give three reasons why it is important to conserve natural habitats and populations.

20)* Here is a graph of population and household waste (destined for landfill) produced for a small village.

   a) In 2002 how many people lived in the village?

   b) In what year did the village produce 12.5 tonnes of waste?

   c) Using your answer from part a), work out how many tonnes one person produced in 2002.

   d) What year did the village bring in a recycling scheme?

   e) Give two ways that a recycling scheme could benefit the environment.

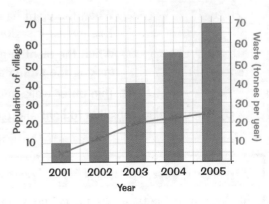

* Answers on page 140

Cells

Ahhh, <u>cells</u>. After all that new-fangled cloning malarky and fretting over the state of the environment, it must come as quite a <u>relief</u> to get back to some good, old-fashioned, boring science. No? Well, <u>tough</u>.

Plant and Animal Cells Have Similarities and Differences

Most <u>human cells</u>, like most <u>animal</u> cells, have the following parts — make sure you know them all:

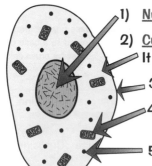

1) <u>Nucleus</u> — contains <u>genetic material</u> that controls the activities of the cell.
2) <u>Cytoplasm</u> — gel-like substance where most of the <u>chemical reactions</u> happen. It contains <u>enzymes</u> (see page 55) that control these chemical reactions.
3) <u>Cell membrane</u> — holds the cell together and controls what goes <u>in</u> and <u>out</u>.
4) <u>Mitochondria</u> — these are where most of the reactions for <u>respiration</u> take place (see page 60). Respiration releases <u>energy</u> that the cell needs to work.
5) <u>Ribosomes</u> — these are where <u>proteins</u> are made in the cell.

Plant cells usually have <u>all the bits</u> that <u>animal</u> cells have, plus a few <u>extra</u> things that animal cells <u>don't</u> have:

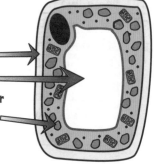

1) Rigid <u>cell wall</u> — made of <u>cellulose</u>. It <u>supports</u> the cell and strengthens it.
2) <u>Permanent vacuole</u> — contains <u>cell sap</u>, a weak solution of sugar and salts.
3) <u>Chloroplasts</u> — these are where <u>photosynthesis</u> occurs, which makes food for the plant (see page 78). They contain a <u>green</u> substance called <u>chlorophyll</u>.

Most Cells are Specialised for Their Function

Similar cells are grouped together to make a <u>tissue</u>, and different tissues work together as an <u>organ</u>. Most cells are <u>specialised</u> for their function within a <u>tissue</u> or <u>organ</u>. In the exam you might have to explain <u>how</u> a particular cell is adapted for its function. Here are a couple of plant examples:

1) Palisade Leaf Cells are Adapted for Photosynthesis

1) They're packed with <u>chloroplasts</u> for <u>photosynthesis</u>.
2) Their <u>tall</u> shape means a lot of <u>surface area</u> is exposed down the side for <u>absorbing CO_2</u> from the air in the leaf.
3) They're <u>thin</u>, so you can pack loads of them in at the top of a leaf.

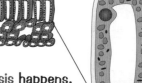

Lots of palisade cells make up <u>palisade tissue</u> where most of the <u>photosynthesis</u> happens.

2) Guard Cells are Adapted to Open and Close Pores

1) They have a special kidney shape which <u>opens</u> and <u>closes</u> pores (<u>stomata</u>) in a leaf.
2) When the plant has <u>plenty</u> of water the guard cells fill with it and go plump (<u>turgid</u>). This makes the stomata <u>open</u> so <u>gases</u> can be exchanged for <u>photosynthesis</u>.
3) When the plant is <u>short</u> of water, the guard cells lose water and go floppy (<u>flaccid</u>), making the stomata <u>close</u>. This helps stop too much water vapour <u>escaping</u>.
4) <u>Thin</u> outer walls and <u>thickened</u> inner walls make the opening and closing work.
5) They're also <u>sensitive to light</u> and <u>close at night</u> to save water without losing out on photosynthesis.

See also red and white blood cells (page 63) and sperm (page 71).

There's quite a bit to learn in biology — but that's life, I guess...

At the top of the page are <u>typical cells</u> with all the typical bits you need to know. But cells <u>aren't</u> all the same — they have different <u>structures</u> and <u>produce</u> different substances depending on the <u>job</u> they do.

DNA

DNA, the molecule of <u>life</u>. Trouble is, it's so darn complicated.

DNA is a Double Helix of Paired Bases

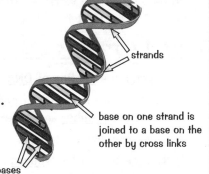

DNA stands for <u>d</u>eoxyribose <u>n</u>ucleic <u>a</u>cid. It contains all the <u>instructions</u> to put an organism together and <u>make it work</u>. It's found in the <u>nucleus</u> of animal and plant cells in <u>long molecules</u> called <u>chromosomes</u> (see p37).

1) A DNA molecule has <u>two strands</u> coiled together in the shape of a <u>double helix</u> (two spirals), as shown in the diagram opposite.

2) Each strand is made up of lots of small groups called "<u>nucleotides</u>".

3) Each <u>nucleotide</u> contains a small molecule called a "<u>base</u>". DNA has just <u>four</u> different bases (shown in the diagram as different colours) — <u>adenine</u> (A), <u>cytosine</u> (C), <u>guanine</u> (G) and <u>thymine</u> (T).

4) The bases are <u>paired</u>, and they always pair up in the same way — it's always A-T and C-G. This is called <u>complementary base-pairing</u>.

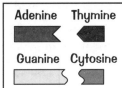

DNA Can Replicate Itself

1) DNA <u>copies itself</u> every time a cell <u>divides</u>, so that each new cell still has the full amount of DNA.

2) In order to copy itself, the DNA double helix first 'unzips' — to form two single <u>strands</u>.

3) As the DNA unwinds itself, <u>new nucleotides</u> (floating about freely in the nucleus) join on <u>only where the bases fit</u> (A with T and C with G), making an <u>exact copy</u> of the DNA on the other strand.

4) The result is <u>two</u> molecules of DNA <u>identical</u> to the original molecule of DNA.

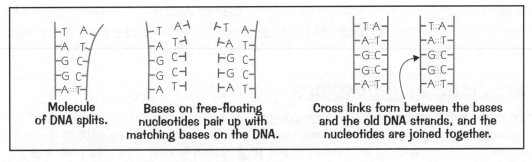

| Molecule of DNA splits. | Bases on free-floating nucleotides pair up with matching bases on the DNA. | Cross links form between the bases and the old DNA strands, and the nucleotides are joined together. |

Everyone has Unique DNA... ...except identical twins and clones

DNA fingerprinting (or genetic fingerprinting) is a way of <u>cutting up</u> a person's DNA into small sections and then <u>separating</u> them. Every person's genetic fingerprint has a <u>unique</u> pattern (unless they're identical twins or clones of course). This means you can <u>tell people apart</u> by <u>comparing samples</u> of their DNA.

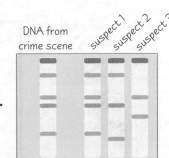

DNA fingerprinting is used in...

1) <u>Forensic science</u> — DNA (from hair, skin flakes, blood, semen etc.) taken from a <u>crime scene</u> is compared with a DNA sample taken from a suspect. In the diagram, suspect 1's DNA has the same pattern as the DNA from the crime scene — so suspect 1 was probably at the crime scene.

2) <u>Paternity testing</u> — to see if a man is the father of a particular child.

So the trick is — frame your twin and they'll never get you...

Some people want there to be a national <u>genetic database</u> of everyone in the country. Then any DNA from a crime scene could easily be identified. But others think this is a big <u>invasion of privacy</u>, and they worry about how <u>safe</u> the data would be and what <u>else</u> it might be used for. There are also <u>scientific problems</u> — false positives can occur if <u>errors</u> are made in the procedure or if the data is <u>misinterpreted</u>.

Making Proteins

So here's how life works — DNA molecules contain a genetic code which determines which proteins are built. The proteins include enzymes that control all the reactions going on in the body. Simple, eh.

DNA Controls the Production of Proteins in a Cell

1) A gene is a section of DNA that 'codes' for a particular protein.

2) Proteins are made up of chains of molecules called amino acids. Each different protein has its own particular number and order of amino acids.

3) This gives each protein a different shape, which means each protein can have a different function.

4) It's the order of the bases in a strand of DNA that decides the order of amino acids in a protein.

5) Each amino acid is coded for by a sequence of three bases in the strand of DNA.

6) Proteins are made from 20 different amino acids, all found in the cytoplasm of cells. They're stuck together to make proteins, following the order of the code in the DNA.

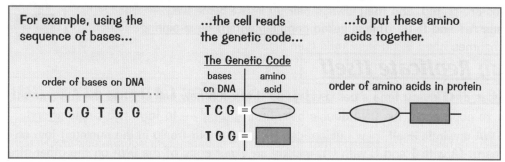

7) We get amino acids from our diet. If we don't take in all the amino acids in the right amounts, our body can change some of them into others. This is called transamination and it happens in the liver.

8) DNA also determines which genes are switched on or off — and so which proteins the cell produces, e.g. haemoglobin or keratin. That in turn determines what type of cell it is, e.g. red blood cell, skin cell.

Proteins are Made by Ribosomes

Proteins are made in the cell by organelles called ribosomes. DNA is found in the cell nucleus and can't move out of it because it's really big. The cell needs to get the information from the DNA to the ribosome in the cell cytoplasm. This is done using a molecule called RNA, which is very similar to DNA, but it's much shorter and only a single strand. RNA is like a messenger between the DNA in the nucleus and the ribosome. Here's how it's done:

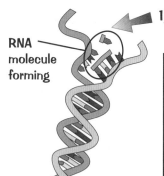

RNA molecule forming

1) The two DNA strands unzip. A molecule of RNA is made using the DNA as a template. Base pairing ensures it's an exact match.

2) The RNA molecule moves out of the nucleus and joins with a ribosome.

3) The job of the ribosome is to stick amino acids together in a chain to make a polypeptide (protein), following the order of bases in the RNA.

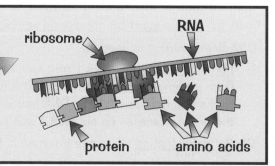

Code of Honour: I shalt not forget this page...

So, the order of bases in your DNA decides what amino acids get joined together, and the order of the amino acids decides the type of protein. And proteins are pretty essential things — all your body's enzymes (see page 55) are proteins, and enzymes control the making of your other, non-protein bits.

Enzymes

Enzymes are Catalysts Produced by Living Things

1) Living things have thousands of different chemical reactions going on inside them all the time.

2) These reactions need to be carefully controlled — to get the right amounts of substances.

3) You can usually make a reaction happen more quickly by raising the temperature. This would speed up the useful reactions but also the unwanted ones too... not good. There's also a limit to how far you can raise the temperature inside a living creature before its cells start getting damaged.

4) So... living things produce enzymes which act as biological catalysts. Enzymes reduce the need for high temperatures and we only have enzymes to speed up the useful chemical reactions in the body.

> A **CATALYST** is a substance which **INCREASES** the speed of a reaction, without being **CHANGED** or **USED UP** in the reaction.

5) Enzymes are all proteins, which is one reason why proteins are so important to living things.

6) All proteins are made up of chains of amino acids. These chains are folded into unique shapes, which enzymes need to do their jobs (see below).

Enzymes Have Special Shapes So They Can Catalyse Reactions

1) Chemical reactions usually involve things either being split apart or joined together.

2) Every enzyme has a unique shape that fits onto the substance or substances involved in a reaction.

3) Enzymes are really picky — they usually only catalyse one reaction.

4) This is because, for the enzyme to work, each substance has to fit its special shape.
If a substance doesn't match the enzyme's shape, then the reaction won't be catalysed.

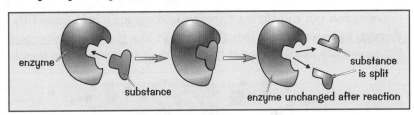

Enzymes Need the Right Temperature and pH

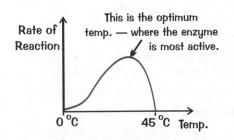

1) Changing the temperature changes the rate of an enzyme-catalysed reaction.

2) Like with any reaction, a higher temperature increases the rate at first. But if it gets too hot, some of the bonds holding the enzyme together break. This destroys the enzyme's special shape and so it won't work any more. It's said to be denatured.

3) Enzymes in the human body normally work best at around 37 °C — body temperature.

4) The pH also affects enzymes. If it's too high or too low, the pH interferes with the bonds holding the enzyme together. This changes the shape and denatures the enzyme.

5) All enzymes have an optimum pH that they work best at. It's often neutral pH 7, but not always — e.g. pepsin is an enzyme used to break down proteins in the stomach. It works best at pH 2, which means it's well-suited to the acidic conditions there.

If only enzymes could speed up revision...

Scientists have caught on to the idea that enzymes are really useful. They're used in biological detergents (to break down nasty stains) and in some baby foods (to predigest the food).

Diffusion

Particles <u>move about randomly</u>, and after a bit they end up <u>evenly spaced</u>. It's not rocket science, is it...

Don't be Put Off by the Fancy Word

"<u>Diffusion</u>" is simple. It's just the <u>gradual movement</u> of particles from places where there are <u>lots</u> of them to places where there are <u>fewer</u> of them. That's all it is — just the <u>natural tendency</u> for stuff to <u>spread out</u>. Unfortunately you also have to learn the fancy way of saying the same thing, which is this:

> **<u>DIFFUSION</u> is the <u>passive movement</u> of <u>particles</u> from an area of <u>HIGHER CONCENTRATION</u> to an area of <u>LOWER CONCENTRATION</u>**

Diffusion happens in both <u>liquids</u> and <u>gases</u> — that's because the particles in these substances are free to <u>move about</u> randomly. The <u>simplest type</u> is when different <u>gases</u> diffuse through each other. This is what's happening when the smell of perfume diffuses through a room:

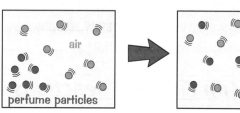

The <u>bigger</u> the <u>difference</u> in concentration, the <u>faster</u> the diffusion rate.

smell diffused in the air

Cell Membranes are Pretty Clever...

They're clever because they <u>hold</u> the cell together <u>but</u> they let stuff <u>in and out</u> as well. Substances can move in and out of cells by <u>diffusion</u> and <u>osmosis</u> (see page 58). Only very <u>small</u> molecules can <u>diffuse</u> through cell membranes though — things like <u>glucose</u>, <u>amino acids</u>, <u>water</u> and <u>oxygen</u>. <u>Big</u> molecules like <u>starch</u> and <u>proteins</u> can't fit through the membrane.

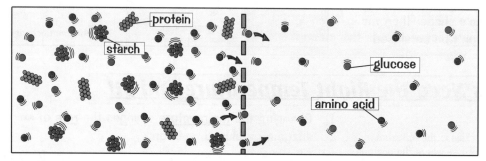

1) Just like with diffusion in air, particles flow through the cell membrane from where there's a <u>higher concentration</u> (more of them) to where there's a <u>lower concentration</u> (not such a lot of them).

2) They're only moving about <u>randomly</u> of course, so they go <u>both</u> ways — but if there are a lot <u>more</u> particles on one side of the membrane, there's a <u>net</u> (overall) movement <u>from</u> that side.

3) The <u>rate</u> of diffusion depends on three main things:

   a) <u>Distance</u> — substances diffuse <u>more quickly</u> when they haven't as <u>far</u> to move. Pretty obvious.

   b) <u>Concentration difference</u> (<u>gradient</u>) — substances diffuse faster if there's a <u>big difference</u> in concentration. If there are <u>lots more</u> particles on one side, there are more there to move across.

   c) <u>Surface area</u> — the <u>more surface</u> there is available for molecules to move across, the <u>faster</u> they can get from one side to the other.

Revision by diffusion — you wish...

Wouldn't that be great — if all the ideas in this book would just gradually drift across into your mind, from an area of <u>high concentration</u> (in the book) to an area of <u>low concentration</u> (in your mind — no offence). Actually, that probably will happen if you read it again. Why don't you give it a go...

Diffusion in Cells

You need to know some <u>examples</u> of <u>where</u> diffusion happens in our bodies.

Alveoli Carry Out Gas Exchange in the Body

1) The <u>lungs</u> contain millions and millions of little air sacs called <u>alveoli</u> where <u>gas exchange</u> happens.

2) The <u>blood</u> passing next to the alveoli has just returned to the lungs from the rest of the body, so it contains <u>lots</u> of <u>carbon dioxide</u> and <u>very little</u> <u>oxygen</u>. <u>Oxygen</u> diffuses <u>out</u> of the <u>alveolus</u> (high concentration) into the <u>blood</u> (low concentration). <u>Carbon dioxide</u> diffuses <u>out</u> of the <u>blood</u> (high concentration) into the <u>alveolus</u> (low concentration) to be breathed out.

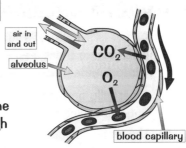

3) When the blood reaches body cells <u>oxygen</u> is released from the <u>red blood cells</u> (where there's a high concentration) and diffuses into the <u>body cells</u> (where the concentration is low).

4) At the same time, <u>carbon dioxide</u> diffuses out of the <u>body cells</u> (where there's a high concentration) into the <u>blood</u> (where there's a low concentration). It's then carried back to the <u>lungs</u>.

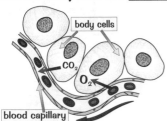

Small Food Molecules Can Diffuse into the Blood

1) Food is <u>digested</u> in the gut to break it down into pieces small enough to be absorbed into the <u>blood</u> by <u>diffusion</u>.

2) The absorption happens in the <u>small intestine</u>, after big molecules like <u>starch</u> and <u>proteins</u> have been broken down into small ones like <u>glucose</u> and <u>amino acids</u>.

3) These molecules can diffuse into the blood from the small intestine because their <u>concentration</u> is <u>higher</u> than it is in the blood.

4) When the blood reaches cells that need these substances because their concentration is <u>low</u>, they can diffuse out easily from an area of <u>higher concentration</u> to an area of <u>lower</u>.

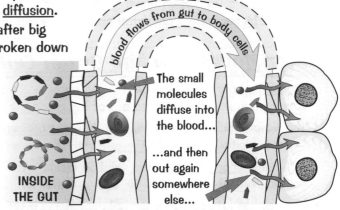

And Finally, Diffusion Happens in Synapses

<u>Neurones</u> (nerve cells) are connected by <u>synapses</u>. A synapse is just a <u>gap</u> between the end of one neurone and the start of the next.

1) When a <u>nerve impulse</u> arrives at a synapse it triggers the release of a <u>transmitter substance</u> from the end of the neurone into the gap.

2) The transmitter substance <u>diffuses</u> across the gap between the neurones and <u>binds</u> to <u>receptors</u> on the end of the <u>next</u> neurone.

3) This stimulates a <u>new nerve impulse</u> in this neurone. The diffusion of the transmitter substance across the synapse allows the nerve impulse to <u>jump the gap</u> and continue on the other side.

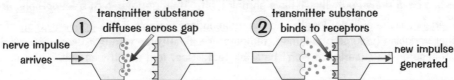

Don't worry — there's still diffusion in a leaf to look forward to...

...just in case you were getting upset at the thought of not hearing any more about it. So that's <u>three</u> <u>examples</u> (and the leaf, see p80). The process of diffusion is the <u>same</u> in each case.

Osmosis

If you've got your head round <u>diffusion</u>, osmosis will be a <u>breeze</u>. If not, what are you doing turning over?

Osmosis *is a Special Case* of *Diffusion*, *That's All*

> <u>OSMOSIS</u> is the <u>movement of water molecules</u> across a <u>partially permeable membrane</u> from a region of <u>higher water concentration</u> to a region of <u>lower water concentration</u>.

1) A <u>partially permeable</u> membrane is just one with very small holes in it. So small, in fact, only tiny <u>molecules</u> (like water) can pass through them, and bigger molecules (e.g. <u>sucrose</u>) can't.

2) The water molecules actually pass <u>both ways</u> through the membrane during osmosis. This happens because water molecules <u>move about randomly</u> all the time.

3) But because there are <u>more</u> water molecules on one side than on the other, there's a steady <u>net flow</u> of water into the region with <u>fewer</u> water molecules, e.g. into the <u>stronger</u> sugar solution.

4) This means the <u>strong</u> solution gets more <u>dilute</u>. The water acts like it's trying to "<u>even up</u>" the concentration either side of the membrane.

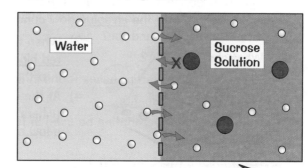

Net movement of water molecules

5) Osmosis is a type of <u>diffusion</u> — passive movement of <u>water particles</u> from an area of <u>higher water concentration</u> to an area of <u>lower water concentration</u>.

Water Moves Into and Out of Cells by Osmosis

1) <u>Tissue fluid</u> surrounds the cells in the body — it's basically just <u>water</u> with <u>oxygen</u>, <u>glucose</u> and stuff dissolved in it. It's squeezed out of the <u>blood capillaries</u> to supply the cells with everything they need.

2) The tissue fluid will usually have a <u>different concentration</u> to the fluid <u>inside</u> a cell. This means that water will either move <u>into the cell</u> from the tissue fluid, or <u>out of the cell</u>, by <u>osmosis</u>.

3) If a cell is <u>short of water</u>, the solution inside it will become quite <u>concentrated</u>. This usually means the solution <u>outside</u> is more <u>dilute</u>, and so water will move <u>into</u> the cell by osmosis.

4) If a cell has <u>lots of water</u>, the solution inside it will be <u>more dilute</u>, and water will be <u>drawn out</u> of the cell and into the fluid outside by osmosis.

There's a fairly dull <u>experiment</u> you can do to show osmosis at work.

You cut up an innocent <u>potato</u> into identical cylinders, and get some beakers with <u>different sugar solutions</u> in them. One should be <u>pure water</u>, another should be a <u>very concentrated sugar solution</u>. Then you can have a few others with concentrations <u>in between</u>.

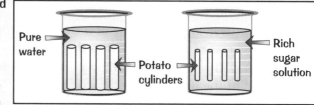

You measure the <u>length</u> of the cylinders, then leave a few cylinders in each beaker for half an hour or so. Then you take them out and measure their lengths <u>again</u>. If the cylinders have drawn in water by osmosis, they'll be a bit <u>longer</u>. If water has been drawn out, they'll have <u>shrunk</u> a bit. Then you can plot a few <u>graphs</u> and things.

The <u>dependent variable</u> is the <u>cylinder length</u> and the <u>independent variable</u> is the <u>concentration</u> of the sugar solution. All <u>other</u> variables (volume of solution, temperature, time, type of sugar used, etc. etc.) must be kept the <u>same</u> in each case or the experiment won't be a <u>fair test</u>. See, told you it was dull.

And to all you cold-hearted potato murderers...

And that's why it's bad to drink sea water. The high <u>salt</u> content means you end up with a much <u>lower water concentration</u> in your blood and tissue fluid than in your cells. Lots of water is sucked out of your cells by osmosis and they <u>shrivel and die</u>. So next time you're stranded at sea, remember this page...

Revision Summary for Section Six

Well, just look at this — it's another Revision Summary page. It's amazing the way they turn up at the end of every section, and they're always a right bundle of laughs, aren't they? I mean, just look at question one there — "Where in the cell does respiration happen?" HAAAAH HAR HAR HAR. Good one.

1) Where in the cell does respiration happen?
2) Name five parts of a cell that both plants and animal cells have.
   What three things do plant cells have that animal cells don't?
3) Give three ways that a palisade leaf cell is adapted for photosynthesis.
4) Give three ways that guard cells are adapted to open and close pores.
5) What does DNA stand for?
6) What shape is DNA?
7) Name the four different bases found in DNA. How do they pair up?
8) Describe DNA replication.
9) How can DNA fingerprinting be used in forensic science?
10) What is a gene?
11) What does a triplet of three bases on a strand of DNA code for?
12) What is transamination?
13) Describe the stages of protein synthesis.
14) What name is given to biological catalysts?
15) What is a catalyst?
16) Give one reason why proteins are so important to living things.
17) What happens at the active site of an enzyme?
18) An enzyme with an optimum temperature of 37 °C is heated to 60 °C.
    Suggest what will happen to it.
19)* The graph on the right shows how the rate of an enzyme-catalysed reaction depends on pH: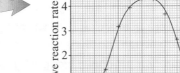
    a) State the optimum pH of the enzyme.
    b) In which part of the human digestive system would you find this enzyme?
20) What is diffusion?
21) What three things does the rate of diffusion depend on?
22) Why does oxygen enter the blood in the alveoli, and leave it when it reaches a respiring tissue?
23) Describe how a nerve impulse travels across a synapse.
24) What is osmosis?
25) A solution of pure water is separated from a concentrated sugar solution by a partially permeable membrane. In which direction will molecules flow, and what substance will these molecules be?
26) An osmosis experiment involves placing pieces of potato into sugar solutions of various concentrations and measuring their lengths before and after. What is:
    a) the independent variable,   b) the dependent variable?

* Answers on page 140

Respiration and Exercise

Respiration happens in little tiny structures called mitochondria (see page 52).

Respiration is NOT 'Breathing In and Out'

Respiration is really important — it releases the energy that cells need to do just about everything.

1) Respiration is the process of breaking down glucose to release energy, and it goes on in every cell in your body. (Glucose contains energy in the form of chemical bonds.)

2) Respiration happens in plants too. All living things respire. It's how they get energy from their food.

This energy is used to do things like:
• build up larger molecules (like proteins)
• contract muscles
• maintain a steady body temperature

RESPIRATION is the process of BREAKING DOWN GLUCOSE TO RELEASE ENERGY, which goes on IN EVERY CELL.

Respiration can be Aerobic or Anaerobic

Aerobic respiration is respiration using oxygen ('aerobic' just means 'with air'). It's the most efficient way to release energy from glucose. Learn the word equation:

glucose + oxygen → carbon dioxide + water (+ ENERGY)

Anaerobic respiration happens when there's not enough oxygen available (e.g. when you're exercising hard). Anaerobic just means without air and it's NOT the best way to release energy from glucose — but it's useful in emergencies. The overall word equation is:

glucose → lactic acid (+ ENERGY)

When You Exercise You Respire More

1) When you exercise, your muscles need more energy so you respire more.

2) You need to get more oxygen into the cells. Your breathing rate increases to get more oxygen into the lungs, and your heart rate increases to get this oxygenated blood around the body faster.

3) During really vigorous exercise (like sprinting), your body can't supply enough oxygen to your muscles quickly enough, so they start respiring anaerobically.

4) Anaerobic respiration produces lactic acid, which builds up in your muscles and causes pain. When you stop exercising, you'll have an oxygen debt. You have to keep breathing hard to repay the oxygen that you didn't manage to get to your muscles. The oxygen breaks down the lactic acid.

5) Athletes monitor their heart rate and breathing rate to help with their training.

6) The current UK government recommendation is to exercise for at least 30 minutes, five times a week in order to stay fit and healthy. But the official advice changes all the time — not so long ago the recommendation was 20 minutes, three times a week.

Alveoli are Specialised for Gas Exchange

1) The huge number of microscopic alveoli gives the lungs an enormous surface area.

2) There's a moist lining for gases to dissolve in.

3) The alveoli have very thin walls — only one cell thick, so the gas doesn't have far to diffuse.

4) They have a great blood supply to maintain a high concentration gradient.

5) The walls are permeable — so gases can diffuse across easily.

Oxygen debt — cheap to pay back...

Advice about exercise (and diet) is based on scientific evidence from many different surveys and studies.

Enzymes and Digestion

The enzymes used in <u>respiration</u> work <u>inside cells</u>. Various different enzymes are used in <u>digestion</u> too, but these enzymes are produced by specialised cells and then <u>released</u> into the <u>gut</u> to mix with the food.

Digestive Enzymes Break Down Big Molecules into Smaller Ones

1) <u>Starch</u>, <u>proteins</u> and <u>fats</u> are BIG molecules. They're too big to pass through the walls of the digestive system.

2) <u>Sugars</u>, <u>amino acids</u>, <u>glycerol</u> and <u>fatty acids</u> are much smaller molecules. They can pass easily through the walls of the digestive system.

3) The <u>digestive enzymes</u> break down the BIG molecules into the smaller ones.

Amylase Converts Starch into Simple Sugars

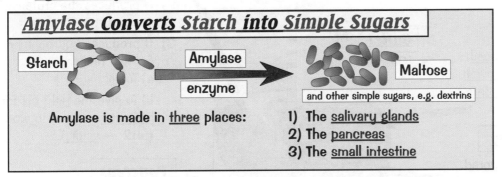

Amylase is made in <u>three</u> places:
  1) The <u>salivary glands</u>
  2) The <u>pancreas</u>
  3) The <u>small intestine</u>

Protease Converts Proteins into Amino Acids

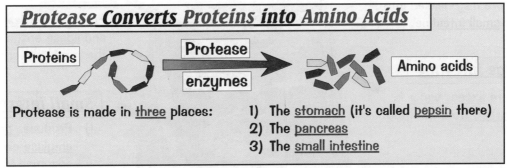

Protease is made in <u>three</u> places:
  1) The <u>stomach</u> (it's called <u>pepsin</u> there)
  2) The <u>pancreas</u>
  3) The <u>small intestine</u>

Lipase Converts Fats into Glycerol and Fatty Acids

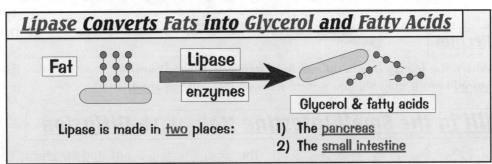

Lipase is made in <u>two</u> places:
  1) The <u>pancreas</u>
  2) The <u>small intestine</u>

Bile Neutralises the Stomach Acid and Emulsifies Fats

1) Bile is <u>produced</u> in the <u>liver</u>. It's <u>stored</u> in the <u>gall bladder</u> before it's released into the <u>small intestine</u>.

2) The <u>hydrochloric acid</u> in the stomach makes the pH <u>too acidic</u> for enzymes in the small intestine to work properly. Bile is <u>alkaline</u> — it <u>neutralises</u> the acid and makes conditions <u>alkaline</u>. The enzymes in the small intestine <u>work best</u> in these alkaline conditions.

3) It <u>emulsifies</u> fats. In other words it breaks the fat into <u>tiny droplets</u>. This gives a much <u>bigger</u> <u>surface area</u> of fat for the enzyme lipase to work on — which makes its digestion <u>faster</u>.

What do you call an acid that's eaten all the pies...

This all happens inside our digestive system, but there are some microorganisms which secrete their digestive enzymes <u>outside their body</u> onto their food. The food's digested, then the microorganism absorbs the nutrients. Nice. I wouldn't like to empty the contents of my stomach onto my plate before eating it.

The Digestive System

So now you know what the enzymes do, here's a nice big picture of the whole of the digestive system.

The Breakdown of Food is Catalysed by Enzymes

Enzymes used in the digestive system are produced by specialised cells in glands and in the gut lining.

Salivary glands

These produce amylase enzyme in the saliva.

Liver

Where bile is produced. Bile neutralises stomach acid and emulsifies fats.

Gall bladder

Where bile is stored, before it's released into the small intestine.

Large intestine

Where excess water is absorbed from the food.

Rectum

Where the faeces (made up mainly of indigestible food) are stored before they bid you a fond farewell through the anus.

Tongue

Gullet
(Oesophagus)

Stomach

1) It pummels the food with its muscular walls.

2) It produces the protease enzyme, pepsin.

3) It produces hydrochloric acid for two reasons:
   a) To kill bacteria
   b) To give the right pH for the protease enzyme to work (pH2 — acidic).

Pancreas

Produces protease, amylase and lipase enzymes. It releases these into the small intestine.

Small intestine

1) Produces protease, amylase and lipase enzymes to complete digestion.

2) This is also where the "food" is absorbed out of the digestive system into the body.

Villi in the Small Intestine Help with Diffusion

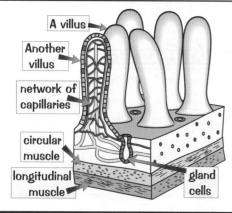

A villus
Another villus
network of capillaries
circular muscle
longitudinal muscle
gland cells

1) The small intestine is adapted for absorption of food.

2) It's very long, so there's time to break down and absorb all the food before it reaches the end.

3) There's a really big surface area for absorption, because the walls of the small intestine are covered in millions and millions of tiny little projections called villi.

4) Each cell on the surface of a villus also has its own microvilli — little projections that increase the surface area even more.

5) Villi have a single permeable layer of surface cells and a very good blood supply to assist quick absorption.

You don't have to bust a gut to revise this page...

Did you know that the whole of your digestive system is actually a hole that goes right through your body? Think about it. It just gets loads of food, digestive juices and enzymes piled into it.

Functions of the Blood

Blood is very useful stuff. It's a big transport system for moving things around the body. The <u>blood cells</u> do good work too. The <u>red blood cells</u> are responsible for transporting <u>oxygen</u> about, and they carry 100 times more than could be moved just dissolved in the plasma. And as for the white blood cells...

Plasma *is the* Liquid Bit *of Blood*

It's basically blood minus the blood cells (see below). Plasma is a pale yellow liquid which <u>carries just about everything</u> that needs transporting around your body:

1) <u>Red</u> and <u>white blood cells</u> (see below) and <u>platelets (used in clotting)</u>.
2) <u>Water</u>.
3) Digested food products like <u>glucose</u> and <u>amino acids</u> from the gut to all the body cells.
4) <u>Carbon dioxide</u> from the body cells to the lungs.
5) <u>Urea</u> from the liver to the kidneys (where it's removed in the urine).
6) <u>Hormones</u> — these act like chemical messengers.
7) <u>Antibodies</u> and <u>antitoxins</u> produced by the white blood cells (see below).

Red Blood Cells *Have the Job of Carrying Oxygen*

They transport <u>oxygen</u> from the <u>lungs</u> to <u>all</u> the cells in the body.
The <u>structure</u> of a red blood cell is adapted to its <u>function</u>:

1) Red blood cells are <u>small</u> and have a <u>biconcave shape</u> (which is a posh way of saying they look a little bit like doughnuts, see diagram below) to give a <u>large surface area</u> for <u>absorbing</u> and <u>releasing oxygen</u>.

2) They contain <u>haemoglobin</u>, which is what gives blood its <u>colour</u> — it contains a lot of <u>iron</u>. In the lungs, haemoglobin <u>reacts with oxygen</u> to become <u>oxyhaemoglobin</u>. In body tissues the reverse reaction happens to <u>release oxygen to the cells</u>.

3) Red blood cells don't have a <u>nucleus</u> — this frees up <u>space</u> for more haemoglobin, so they can carry more oxygen.

4) Red blood cells are very <u>flexible</u>. This means they can easily pass through the <u>tiny capillaries</u> (see next page).

White Blood Cells *are Used to Fight Disease*

1) Their main role is <u>defence against disease</u>.
2) They produce <u>antibodies</u> to fight microbes.
3) They produce <u>antitoxins</u> to neutralise the toxins produced by microbes.
4) They have a <u>flexible shape</u>, which helps them to <u>engulf</u> any microorganisms they come across inside the body. Basically the white blood cell wraps around the microorganism until it's <u>totally surrounded</u>, and then it <u>digests it</u> using enzymes.

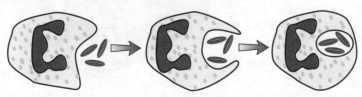

What do white blood cells and elephants have in common...

The average adult human body contains about <u>five and a half litres</u> of blood altogether, and every single drop contains <u>millions</u> of cells. There are usually about 500 times more red blood cells than white.

Blood Vessels

Blood needs a good system to move it around the body — called the <u>circulatory system</u>.

Blood Vessels **are** Designed **for Their** Function

There are <u>three</u> different types of <u>blood vessel</u>:

1) <u>ARTERIES</u> — these carry the blood <u>away</u> from the heart.
2) <u>CAPILLARIES</u> — these are involved in the <u>exchange of materials</u> at the tissues.
3) <u>VEINS</u> — these carry the blood <u>to</u> the heart.

Arteries **Carry Blood Under** Pressure

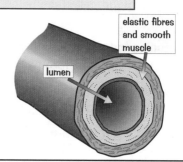

1) The heart pumps the blood out at <u>high pressure</u> so the artery walls are <u>strong</u> and <u>elastic</u>.

2) The walls are <u>thick</u> compared to the size of the hole down the middle (the "<u>lumen</u>" — silly name). They contain thick layers of <u>muscle</u> to make them <u>strong</u>.

<u>Cholesterol</u> is a <u>fatty</u> substance. Eating a diet high in <u>saturated fat</u> has been linked to high levels of cholesterol in the blood. You need some cholesterol for things like <u>making cell membranes</u>. But if you get <u>too much</u> cholesterol it starts to <u>build up</u> in your <u>arteries</u>. These form <u>plaques</u> in the wall of the <u>lumen</u>, which <u>narrows</u> the artery. This <u>restricts</u> the flow of blood — <u>bad news</u> for the part of the body the artery is supplying with <u>food</u> and <u>oxygen</u>. If an artery supplying the <u>heart</u> or <u>brain</u> is affected, it can cause a <u>heart attack</u> or <u>stroke</u>.

Capillaries **are Really Small**

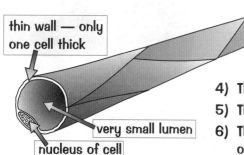

1) Arteries branch into <u>capillaries</u>.

2) Capillaries are really <u>tiny</u> — too small to see.

3) They carry the blood <u>really close</u> to <u>every cell</u> in the body to <u>exchange substances</u> with them.

4) They have <u>permeable</u> walls, so substances can <u>diffuse</u> in and out.

5) They supply <u>food</u> and <u>oxygen</u>, and take away <u>wastes</u> like CO_2.

6) Their walls are usually <u>only one cell thick</u>. This <u>increases</u> the rate of diffusion by <u>decreasing</u> the <u>distance</u> over which it happens.

Veins **Take Blood Back to the Heart**

1) Capillaries eventually <u>join up</u> to form <u>veins</u>.

2) The blood is at <u>lower pressure</u> in the veins so the walls don't need to be as <u>thick</u> as artery walls.

3) They have a <u>bigger lumen</u> than arteries to help the blood <u>flow</u> despite the lower pressure.

4) They also have <u>valves</u> to help keep the blood flowing in the <u>right direction</u>.

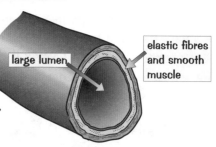  

Learn this page — don't struggle in vein...

Here's an interesting fact for you — your body contains about <u>60 000 miles</u> of blood vessels. That's about <u>six times</u> the distance from <u>London</u> to <u>Sydney</u> in Australia. Of course, capillaries are really tiny, which is how there can be such a big length — they can only be seen with a <u>microscope</u>.

The Heart

Blood doesn't just move around the body <u>on its own</u>, of course. It needs a <u>pump</u>.

Mammals Have a Double Circulatory System

1) The first one connects the <u>heart</u> to the <u>lungs</u>. <u>Deoxygenated</u> blood is pumped to the <u>lungs</u> to take in <u>oxygen</u>. The blood then <u>returns</u> to the heart.

2) The second one connects the <u>heart</u> to the <u>rest of the body</u>. The <u>oxygenated</u> blood in the heart is pumped out to the <u>body</u>. It <u>gives up</u> its oxygen, and then the <u>deoxygenated</u> blood <u>returns</u> to the heart to be pumped out to the <u>lungs</u> again.

3) Returning the blood to the <u>heart</u> after it's picked up oxygen at the <u>lungs</u> means it can be pumped out around the body with <u>much greater force</u>. This is needed so the blood can get to <u>every last tissue</u> in the body and <u>still</u> have enough push left to flow <u>back to the heart</u> through the veins.

Lungs

Rest of Body

Learn This Diagram of the Heart with All Its Labels

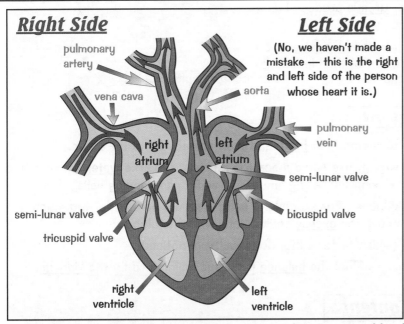

Right Side

pulmonary artery

vena cava

right atrium

semi-lunar valve

tricuspid valve

right ventricle

Left Side

(No, we haven't made a mistake — this is the right and left side of the person whose heart it is.)

aorta

pulmonary vein

semi-lunar valve

bicuspid valve

left atrium

left ventricle

1) The <u>right atrium</u> of the heart receives <u>deoxygenated</u> blood from the <u>body</u> (through the <u>vena cava</u>).
(The plural of atrium is atria.)

2) The deoxygenated blood moves through to the <u>right ventricle</u>, which pumps it to the <u>lungs</u> (via the <u>pulmonary artery</u>).

3) The <u>left atrium</u> receives <u>oxygenated</u> blood from the <u>lungs</u> (through the <u>pulmonary vein</u>).

4) The oxygenated blood then moves through to the <u>left ventricle</u>, which pumps it out round the <u>whole body</u> (via the <u>aorta</u>).

5) The <u>left</u> ventricle has a much <u>thicker wall</u> than the <u>right</u>

ventricle. It needs more <u>muscle</u> because it has to pump blood around the <u>whole body</u>, whereas the right ventricle only has to pump it to the <u>lungs</u>. The <u>valves</u> prevent the <u>backflow</u> of blood.

If the Heart Stops Working Properly — Bits Can be Replaced

The heart has a <u>pacemaker</u> — a group of cells which determine <u>how fast</u> it beats. If this stops working the heartbeat becomes <u>irregular</u>, which can be dangerous. The pacemaker can be <u>replaced</u> with an <u>artificial</u> one. Defective <u>heart valves</u> can also be replaced — either with <u>animal</u> or <u>mechanical</u> valves.

In extreme cases, the <u>whole heart</u> can be <u>removed</u> and <u>replaced</u> by another one from a <u>human donor</u> — this is called a <u>transplant</u>. It involves <u>major surgery</u> and a lifetime of <u>drugs</u> and <u>medical care</u>. They're only done on patients whose hearts are so damaged that the problems <u>can't</u> be solved in any other way. The new heart must be the <u>right size</u>, <u>relatively young</u> and a <u>close tissue match</u> to prevent rejection:

TRANSPLANTS CAN BE REJECTED One of the main problems with heart transplants is that the patient's <u>immune system</u> often recognises the new heart as '<u>foreign</u>' and <u>attacks</u> it — this is called <u>rejection</u>. Doctors use <u>drugs</u> that <u>suppress</u> the patient's immune system to help <u>stop</u> the donor heart being rejected, but that leaves the patient more <u>vulnerable</u> to <u>infections</u>.

Okay — let's get to the heart of the matter...

The human heart <u>beats</u> 100 000 times a day on average. You can measure it by taking your <u>pulse</u>.

The Kidneys and Homeostasis

The <u>kidneys</u> are really important in homeostasis (see p11-12) — they control the content of the <u>blood</u>.

Kidneys Basically Act as Filters to "Clean the Blood"

The <u>kidneys</u> perform <u>three main roles</u>:

1) <u>Removal of urea</u> from the blood.
2) <u>Adjustment of ions</u> in the blood.
3) <u>Adjustment of water content</u> of the blood.

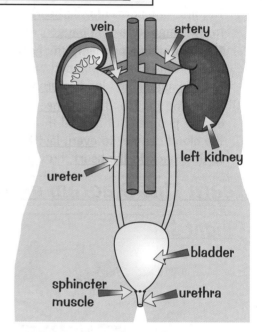

1) Removal of Urea

1) Proteins can't be <u>stored</u> by the body — so any excess amino acids are converted into <u>fats</u> and <u>carbohydrates</u>, which can be stored.

2) This process happens in the <u>liver</u>. <u>Urea</u> is produced as a <u>waste product</u> from the reactions.

3) Urea is <u>poisonous</u>. It's released into the <u>bloodstream</u> by the liver. The <u>kidneys</u> then filter it out of the blood and it's excreted from the body in <u>urine</u>.

2) Adjustment of Ion Content

1) <u>Ions</u> such as <u>sodium</u> are taken into the body in <u>food</u>, and then absorbed into the blood.

2) If the ion content of the body is <u>wrong</u>, this could mean too much or too little <u>water</u> is drawn into cells by <u>osmosis</u> (see page 58). Having the wrong amount of water can <u>damage</u> cells.

3) Excess ions are <u>removed</u> by the kidneys. For example, a salty meal will contain far too much sodium and so the kidneys will remove the <u>excess</u> sodium ions from the blood.

4) Some ions are also lost in <u>sweat</u> (which tastes salty, you may have noticed).

5) But the important thing to remember is that the <u>balance</u> is always maintained by the <u>kidneys</u>.

3) Adjustment of Water Content

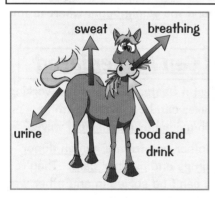

Water is taken into the body as <u>food and drink</u> and is <u>lost</u> from the body in <u>three main ways</u>: 1) In <u>urine</u>
2) In <u>sweat</u>
3) In the air we <u>breathe out</u>.

The body has to <u>constantly balance</u> the water coming in against the water going out. Your body can't control how much you lose in your breath, but it can control the other factors. This means the <u>water balance</u> is between: 1) Liquids <u>consumed</u>
2) Amount <u>sweated out</u>
3) Amount <u>excreted by the kidneys</u> in the <u>urine</u>.

On a <u>cold</u> day, if you <u>don't sweat</u>, you'll produce <u>more urine</u> which will be <u>pale</u> and <u>dilute</u>.

On a <u>hot</u> day, you <u>sweat a lot</u>, and you'll produce <u>less urine</u> which will be <u>dark-coloured</u> and <u>concentrated</u>.

The water lost when it is hot has to be <u>replaced</u> with water from food and drink to restore the <u>balance</u>.

Adjusting water content — blood, sweat and, erm, wee...

Scientists have made a machine which can do the kidneys' job for us — a <u>kidney dialysis machine</u>. People with <u>kidney failure</u> have to use it for 3-4 hours, 3 times a week. Unfortunately it's <u>not</u> something you can carry around in your back pocket, which makes life difficult for people with kidney failure.

The Pancreas and Diabetes

Scientific discoveries often take a <u>long time</u>, and a lot of <u>trial and error</u> — here's a rather famous example.

Insulin was Discovered by Banting and Best

It has been known for some time that people who suffer from diabetes have a lot of <u>sugar</u> in their <u>urine</u>. In the 19th century, scientists <u>removed pancreases</u> from dogs, and the same sugary urine was observed — the dogs became <u>diabetic</u>. That suggested that the pancreas had to have something to do with the illness. In the 1920s Frederick <u>Banting</u> and his assistant Charles <u>Best</u> managed to successfully <u>isolate insulin</u> — the hormone that controls blood sugar levels.

1) Banting and Best <u>tied string</u> around a dog's pancreas so that a lot of the organ <u>wasted</u> away — but the bits which made the <u>hormones</u> were left <u>intact</u>.
2) They <u>removed</u> the pancreas from the dog, and obtained an <u>extract</u> from it.
3) They then injected this extract into <u>diabetic dogs</u> and observed the effects on their <u>blood sugar levels</u>.
4) After the pancreatic extract was <u>injected</u>, the dog's blood sugar level <u>fell dramatically</u>. This showed that the <u>pancreatic extract</u> caused a <u>temporary decrease</u> in <u>blood sugar level</u>.
5) They went on to <u>isolate</u> the substance in the pancreatic extract — <u>insulin</u>.

Diabetes Can be Controlled by Regular Injections of Insulin

After <u>a lot</u> more experiments, Banting and Best tried <u>injecting insulin</u> into a <u>diabetic human</u>. And it <u>worked</u>. Since then insulin has been <u>mass produced</u> to meet the <u>needs</u> of diabetics. Diabetics have to inject themselves with insulin <u>often</u> — 2-4 times a day. They also need to carefully control their <u>diet</u> and the amount of <u>exercise</u> they do (see page 13).

1) At first, the insulin was extracted from the pancreases of <u>pigs</u> or <u>cows</u>. Diabetics used <u>glass syringes</u> that had to be boiled before use.
2) In the 1980s <u>human</u> insulin made by <u>genetic engineering</u> became available. This didn't cause any <u>adverse reactions</u> in patients, which <u>animal</u> insulin sometimes did.
3) <u>Slow</u>, <u>intermediate</u> and <u>fast</u> acting insulins have been developed to make it easier for diabetics to <u>control</u> their blood sugar levels.
4) Ready sterilised, <u>disposable syringes</u> are now available, as well as <u>needle-free devices</u>.

These improved treatment methods allow diabetics to <u>control</u> their blood sugar <u>more easily</u>. This helps them avoid some of the damaging side effects of poor control, such as <u>blindness</u> and <u>gangrene</u>.

Diabetics May Have a Pancreas Transplant

Injecting yourself with insulin every day <u>controls</u> the effects of diabetes, but it doesn't help to <u>cure</u> it.

1) Diabetics can have a <u>pancreas transplant</u>. A successful operation means they won't have to inject themselves with insulin again. But as with any organ transplant, your body can <u>reject</u> the tissue. If this happens you have to take <u>costly immunosuppressive drugs</u>, which often have <u>serious side-effects</u>.
2) Another method, still in its <u>experimental stage</u>, is to transplant just the <u>cells</u> which produce insulin. There's been <u>varying success</u> with this technique, and there are still problems with <u>rejection</u>.
3) Modern research into <u>artificial pancreases</u> and <u>stem cell research</u> may mean the elimination of organ rejection, but there's a way to go yet (see page 42).

Blimey — all that in the last hundred years...

Insulin can't be taken in a pill or tablet — the <u>enzymes</u> in the stomach completely <u>destroy it</u> before it reaches the bloodstream. That's why diabetics have to <u>inject it</u>. Diabetes is becoming more and more <u>common</u>, partly due to our society becoming increasingly overweight. It's very serious.

Revision Summary for Section Seven

And where do you think you're going? It's no use just reading through and thinking you've got it all — this stuff will only stick in your head if you've learnt it properly. And that's what these questions are for. I won't pretend they'll be easy — they're not meant to be, but all the information's in the section somewhere. Have a go at all the questions, then if there are any you can't answer, go back, look stuff up and try again. Enjoy...

1)   Write down the word equations for aerobic respiration and anaerobic respiration.

2)   Give one advantage and one disadvantage of anaerobic respiration.

3)   Give three ways that alveoli are adapted for gaseous exchange.

4)*   Danny measured his heart rate before, during and after exercise. He plotted a graph of the results. Look at the graph and then answer the three questions below.

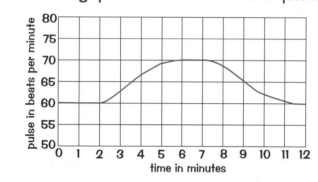

a) What was Danny's heart rate (in beats per minute) when he was at rest?
b) After how many minutes did Danny start exercising?
c) What was Danny's highest heart rate?

5)   How much exercise does the UK government recommend you do per week?

6)   In which three places in the body is amylase produced?

7)   Where in the body is bile: a) produced?   b) stored?   c) used?

8)   Explain why the stomach produces hydrochloric acid.

9)   What is the main function of the small intestine?

10)   Give three ways that the small intestine is adapted for absorption.

11)   Name six things that blood plasma transports around the body.

12)   Name the substance formed in red blood cells when haemoglobin reacts with oxygen.

13)   Why do arteries need very muscular, elastic walls?

14)   Explain how capillaries are adapted to their function.

15)   Name the blood vessel that joins to the right ventricle of the heart. Where does it take the blood?

16)   Why does the left ventricle have a thicker wall than the right ventricle?

17)   What three main jobs do the kidneys do in the body?

18)   Where in the body is urea produced?

19)   What damage could be done in the body if the ion content is wrong?

20)   Explain why your urine is likely to be more concentrated on a hot day.

21)   Describe the experiments by Banting and Best that led to the isolation of insulin.

22)   What are the advantages and disadvantages of a pancreas transplant for a diabetic?

* Answers on page 140

Growth

This topic's about <u>growth and development</u> in plants and animals. Organisms grow using a combination of cell <u>division</u>, cell <u>elongation</u> and cell <u>differentiation</u> — which you'll learn all about in exquisite detail as you go through the topic. But first, here's a bit of general stuff about growth...

Growth *is an Increase in Size or Weight*

You can <u>measure</u> the <u>growth</u> of an organism in these three ways:

1) **Size** — You can measure its <u>height</u>, <u>length</u>, <u>width</u> or <u>circumference</u>.

2) **Wet weight** — Organisms <u>contain</u> a lot of <u>water</u>. The weight of the organism depends on how much water it has gained or lost (e.g. through drinking or sweating). The <u>wet weight</u> of the organism is its weight <u>including all the water</u> in its body — it can vary a lot from <u>one day to the next</u>.

3) **Dry weight** — The <u>dry weight</u> is the weight of an organism with <u>no water in its body</u>. This doesn't vary in the same way as wet weight, but you can only measure it once the organism's dead. The dead organism is <u>dried out</u> by leaving it in a hot oven overnight — then what's left is weighed.

Animals *Stop Growing, Plants Can Grow Continuously*

Plants and animals <u>grow differently</u>:

1) Animals tend to grow while they're <u>young</u>, and then they reach <u>full growth</u> and <u>stop</u> growing. Plants often grow <u>continuously</u> — even really old trees will keep putting out <u>new branches</u>.

2) In animals, growth happens by <u>cell division</u>, but in plants, growth in <u>height</u> is mainly due to <u>cell enlargement</u> (elongation) — cell <u>division</u> usually just happens in the <u>tips</u> of the <u>roots</u> and <u>shoots</u>.

Some *Animals are Able to Regenerate*

1) A few animals have the ability to <u>regrow</u> (regenerate) part of their body if it is <u>damaged</u>:

- If some types of <u>worm</u> are cut in two, the front part can grow a new 'tail'.
- If a <u>young spider</u> loses a leg, it can grow a new one (adult spiders can't, though).
- Some <u>reptiles</u>, like lizards, can regrow a lost leg or tail.

2) The ability to regenerate parts of the body is <u>pretty rare</u> though. It tends to happen in fairly simple (or very young) animals which still contain lots of <u>stem cells</u> (see p.73).

Human Growth *Can be Monitored and Manipulated*

1) Humans, like other mammals, give birth to <u>live young</u>. The baby grows inside the womb until it reaches a stage where it can <u>survive</u> outside — this period is called <u>gestation</u> and it's 38 weeks in humans.

2) We don't grow evenly in the womb or in early life — certain organs grow <u>faster</u> than others, and the <u>fastest-growing</u> of all is the <u>brain</u>. This is because a large and well-developed brain gives humans a big <u>survival advantage</u> — it's our best tool for finding food, avoiding predators, etc.

3) A baby's growth rate can be monitored by measuring its <u>head circumference</u>. The actual values are not as important as the <u>rate</u> of growth. If the baby is growing too <u>slowly</u>, or if the head is relatively too <u>large</u> or <u>small</u>, it can alert the doctor to possible <u>development problems</u>.

4) Growth factors are chemicals that stimulate the body to grow. Some athletes have used <u>growth factor</u> drugs to <u>improve their performance</u> at <u>sport</u>. This is <u>illegal</u> because it gives them an <u>unfair advantage</u> over other competitors. It also has health risks:

- Growth factor drugs can have <u>bad side effects</u> for health: they can <u>reduce fertility</u>, can increase the risk of <u>heart disease</u> and can sometimes trigger mental illnesses like <u>depression</u>.
- Some growth factors cause <u>women</u> to develop <u>male characteristics</u>, e.g. a deeper voice.

Growth — birds do it, bees do it, even educated fleas do it...

I wonder what it'd be like to grow an extra leg. Could help in P.E., but finding clothes might be a problem.

Cell Division — Mitosis

In order to <u>survive</u> and <u>grow</u>, our cells have got to be able to <u>divide</u>. And that means our <u>DNA</u> as well...

Mitosis Makes New Cells for Growth and Repair

<u>Body cells</u> are <u>diploid</u> — they normally have <u>two copies</u> of each <u>chromosome</u>, one from the organism's '<u>mother</u>', and one from its '<u>father</u>'. So, humans have two copies of chromosome 1, two copies of chromosome 2, etc. The diagram shows the <u>23 pairs of chromosomes</u> from a human cell. The 23rd pair is a bit different.

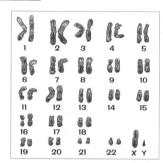

When a body cell <u>divides</u> it needs to make new cells <u>identical</u> to the <u>original</u> cell — with the <u>same number</u> of chromosomes.

This type of cell division is called <u>mitosis</u>. It's used when plants and animals want to <u>grow</u> or to <u>replace</u> cells that have been <u>damaged</u>.

"<u>MITOSIS</u> is when a cell reproduces itself <u>by splitting</u> to form <u>two identical offspring</u>."

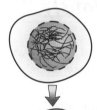

In a cell that's not dividing, the DNA is all spread out in <u>long strings</u>.

If the cell gets a signal to <u>divide</u>, it needs to <u>duplicate</u> its DNA — so there's one copy for each new cell. The DNA is copied and forms <u>X-shaped</u> chromosomes. Each 'arm' of the chromosome is an <u>exact duplicate</u> of the other.

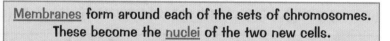

The left arm has the same DNA as the right arm of the chromosome.

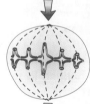

The chromosomes then <u>line up</u> at the centre of the cell and <u>cell fibres</u> pull them apart. The <u>two arms</u> of each chromosome go to <u>opposite ends</u> of the cell.

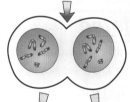

Membranes form around each of the sets of chromosomes. These become the <u>nuclei</u> of the two new cells.

Lastly, the <u>cytoplasm</u> divides.

You now have <u>two new cells</u> containing exactly the same DNA — they're <u>identical</u>.

Most cells have a limit to the number of times they can divide. This is called the <u>Hayflick limit</u>. Stem cells (see page 73) and cancer cells have no limit — that's what makes cancerous cells so dangerous.

Asexual reproduction in some organisms uses mitosis — there's more on that back on page 41.

A cell's favourite computer game — divide and conquer...

This can seem tricky at first. But don't worry — just go through it <u>slowly</u>, one step at a time. This type of division produces identical cells, but there's another type which doesn't... (see next page)

Cell Division — Meiosis

You thought mitosis was exciting. Hah. <u>You ain't seen nothing yet</u>.

Meiosis Involves Two Divisions

<u>Gametes</u> (sperm and egg cells) only have <u>half</u> the number of chromosomes as a normal body cell. They are produced by <u>meiosis</u>. In humans it <u>only</u> happens in the <u>reproductive organs</u> (i.e. ovaries and testes).

| "<u>MEIOSIS</u> produces <u>four haploid</u> cells whose chromosomes are NOT identical." |

chromosome pair

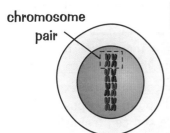

Meiosis — Division 1

1) As with mitosis, before the cell starts to divide, it <u>duplicates</u> its <u>DNA</u> — one arm of each chromosome is an <u>exact copy</u> of the other arm.

2) In the <u>first division</u> in meiosis (there are two divisions) the chromosome pairs (see previous page) <u>line up</u> in the centre of the cell.

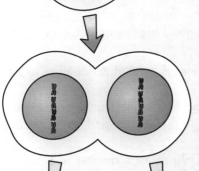

3) They're then <u>pulled apart</u>, so each new cell only has one copy of each chromosome. <u>Some</u> of the father's chromosomes (shown in blue) and <u>some</u> of the mother's chromosomes (shown in red) go into each new cell.

4) Each new cell will have a <u>mixture</u> of the mother's and father's chromosomes. Mixing up the genes in this way creates <u>variation</u> in the offspring. This is a huge <u>advantage</u> of <u>sexual</u> reproduction over <u>asexual</u> reproduction.

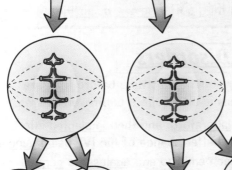

Meiosis — Division 2

5) In the <u>second division</u> the chromosomes <u>line up</u> again in the centre of the cell. It's a lot like mitosis. The arms of the chromosomes are <u>pulled apart</u>.

6) You get <u>four gametes</u>, each with only a <u>single set</u> of chromosomes in it.

After two gametes join at fertilisation, the cell grows by repeatedly dividing by <u>mitosis</u>.

Sperm Cells are Adapted for Their Function

The <u>function</u> of a sperm is to <u>transport</u> the <u>male's DNA</u> to the <u>female's egg</u> so their DNA can <u>combine</u>.

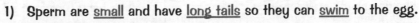

1) Sperm are <u>small</u> and have <u>long tails</u> so they can <u>swim</u> to the egg.

2) Sperm have lots of <u>mitochondria</u> (see page 52) to provide the <u>energy</u> needed to swim this distance.

3) Sperm also have an <u>acrosome</u> at the front of the 'head', where they store the <u>enzymes</u> they need to <u>digest</u> their way through the <u>membrane</u> of the egg cell.

4) They're produced in <u>large numbers</u> to <u>increase</u> the chance of fertilisation.

Relegation to the Second Division is inevitable...

As usual, the best thing to do is to <u>learn the diagram</u>. Cover it up and sketch it out.

Sexual Reproduction — Ethics

During sexual reproduction a sperm cell and an egg cell combine to produce an embryo. If problems with the embryo are detected during it's development the parents can face some difficult decisions...

A Pregnancy Can Legally be Terminated Up to 24 Weeks

After the 8th week of pregnancy, the embryo starts to look a bit more human and is called a foetus. In Britain, a termination (induced abortion) is legal until a foetus is 24 weeks old if two doctors agree that termination is necessary. An abortion can be carried out later than this if the pregnancy is putting the mother's health at serious risk or if there is a major foetal abnormality. The 24-week limit came into effect in 1991, but it remains the subject of some fairly heated debate:

1) Some people argue that abortion at any stage of pregnancy is unethical. They argue that human life starts at fertilisation — and ending pregnancy is the same as killing a human being.

2) Other people argue that the foetus doesn't become human until it's conscious — for example, when it starts feeling pain. They argue that abortion should be allowed up until this point. But it's difficult to pinpoint exactly when the foetus becomes conscious and can feel pain. Some people argue that it's the point when pain receptors first develop at about 7 weeks. Others argue that the foetus can't feel pain until the pain receptors are connected up in the brain — which doesn't happen until about 26 weeks.

3) The legal argument in Britain is based on the 'viability' of the foetus — that is, whether or not the foetus can survive outside the womb (with medical help).
   With advances in medicine, foetuses are becoming viable earlier in the pregnancy — babies have survived from as early as 21 weeks, so some people argue the limit should be dropped from 24 weeks to 20 weeks.
   But babies born so prematurely can have serious problems. Only about a quarter of babies born at 24 weeks or under survive and, of those, over a third suffer severe disabilities.

Embryos Can be Screened for Genetic Disorders

For parents who are undergoing IVF, embryos can be tested for genetic disorders before they are implanted into the mother. But, like terminating a pregnancy this is also quite a controversial area.

1) During in vitro fertilisation (IVF), embryos are fertilised in a laboratory, and then implanted into the mother's womb. More than one egg is fertilised, so there's a better chance of the IVF being successful.
2) Before being implanted, it's possible to remove a cell from each embryo and analyse its genes.
3) Many genetic disorders could be detected in this way, such as cystic fibrosis and Huntington's.
4) Embryos with 'good' genes would be implanted into the mother — the ones with 'bad' genes destroyed.

There is a huge debate raging about embryonic screening. Here are some arguments for and against it.

Against Embryonic Screening	For Embryonic Screening
1) There may come a point where everyone wants to screen their embryos so they can pick the most 'desirable' one, e.g. they may want a blue eyed, blonde haired, intelligent boy.	1) It will help to stop people suffering.
	2) There are laws to stop it going too far. At the moment parents cannot even select the sex of their baby (unless it's for health reasons).
2) The rejected embryos are destroyed — they could have developed into humans.	3) During IVF, most of the embryos are destroyed anyway — screening just allows the selected one to be healthy.
3) It implies that people with genetic problems are 'undesirable' — this could increase prejudice.	4) Treating disorders costs the Government (and the taxpayers) a lot of money.

Termination is an emotional topic...

...and a lot of people have very strong views about it. In the exam you've got to be able to back up your view with a valid argument, and then show that you have considered other points of view.

Stem Cells and Differentiation

Stem cell research has exciting possibilities, but it's also pretty controversial.

Embryonic Stem Cells Can Turn into ANY Type of Cell

1) Most cells in your body are specialised for a particular job. E.g. white blood cells are brilliant at fighting invaders but can't carry oxygen, like red blood cells.

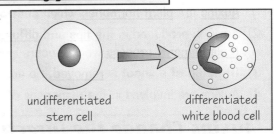

undifferentiated stem cell differentiated white blood cell

2) Differentiation is the process by which a cell changes to become specialised for its job. In most animal cells, the ability to differentiate is lost at an early stage, but lots of plant cells don't ever lose this ability.

3) Some cells are undifferentiated. They can develop into different types of cell depending on what instructions they're given. These cells are called STEM CELLS.

4) Stem cells are found in early human embryos. They're exciting to doctors and medical researchers because they have the potential to turn into any kind of cell at all. This makes sense if you think about it — all the different types of cell found in a human being have to come from those few cells in the early embryo.

5) Adults also have stem cells, but they're only found in certain places, like bone marrow. These aren't as versatile as embryonic stem cells — they can't turn into any cell type at all, only certain ones.

Stem Cells May be Able to Cure Many Diseases

1) Medicine already uses adult stem cells to cure disease. For example, people with some blood diseases (e.g. sickle cell anaemia) can be treated by bone marrow transplants. Bone marrow contains stem cells that can turn into new blood cells to replace the faulty old ones.

2) Scientists can also extract stem cells from very early human embryos and grow them.

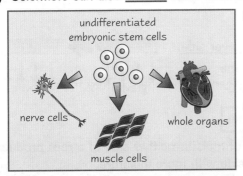

undifferentiated embryonic stem cells

nerve cells whole organs

muscle cells

3) These embryonic stem cells could be used to replace faulty cells in sick people — you could make beating heart muscle cells for people with heart disease, insulin-producing cells for people with diabetes, nerve cells for people paralysed by spinal injuries, and so on.

4) To get cultures of one specific type of cell, researchers try to control the differentiation of the stem cells by changing the environment they're growing in. So far, it's still a bit hit-and-miss — lots more research is needed.

Some People are Against Stem Cell Research

1) Some people are against stem cell research because they feel that human embryos shouldn't be used for experiments since each one is a potential human life.

2) Others think that curing patients who already exist and who are suffering is more important than the rights of embryos.

3) One fairly convincing argument in favour of this point of view is that the embryos used in the research are usually unwanted ones from fertility clinics which, if they weren't used for research, would probably just be destroyed. But of course, campaigners for the rights of embryos usually want this banned too.

4) These campaigners feel that scientists should concentrate more on finding and developing other sources of stem cells, so people could be helped without having to use embryos.

5) In some countries stem cell research is banned, it's allowed in the UK but it must follow strict guidelines.

But florists cell stems, and nobody complains about that...

The potential of stem cells is huge — but it's early days yet. Research has recently been done into getting stem cells from alternative sources. For example, from umbilical cords.

Growth in Plants

Plants <u>don't</u> grow randomly. Plant hormones make sure they grow in a <u>useful direction</u> (e.g. towards light).

Auxins are Plant Growth Hormones

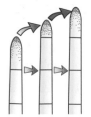

1) <u>Auxins</u> are <u>plant hormones</u> which control <u>growth</u> near the <u>tips</u> of <u>shoots</u> and <u>roots</u>.
2) Auxin is produced in the <u>tips</u> and <u>diffuses backwards</u> to stimulate the <u>cell elongation (enlargement) process</u> which occurs in the cells <u>just behind</u> the tips.
3) If the tip of a shoot is <u>removed</u>, no auxin is available and the shoot may <u>stop growing</u>.
4) Auxins are involved in the responses of plants to <u>light</u>, <u>gravity</u> and <u>water</u>.

Auxins Change the Direction of Root and Shoot Growth

Extra auxin <u>promotes</u> growth in the <u>shoot</u> but actually <u>inhibits</u> growth in the <u>root</u> — but this produces the <u>desired result</u> in <u>both cases</u>.

Shoots grow towards light

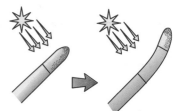

1) When a <u>shoot tip</u> is exposed to <u>light</u>, <u>more auxin</u> accumulates on the side that's in the <u>shade</u> than the side that's in the light.
2) This makes the cells grow (elongate) <u>faster</u> on the <u>shaded side</u>, so the shoot bends <u>towards</u> the light.

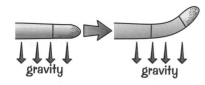

Shoots grow away from gravity

1) When a <u>shoot</u> is growing sideways, <u>gravity</u> produces an unequal distribution of auxin in the tip, with <u>more auxin</u> on the <u>lower side</u>.
2) This causes the lower side to grow <u>faster</u>, bending the shoot <u>upwards</u>.

Roots grow towards gravity

1) A <u>root</u> growing sideways will also have more auxin on its <u>lower side</u>.
2) But in a root the <u>extra</u> auxin <u>inhibits</u> growth. This means the cells on <u>top</u> elongate faster, and the root bends <u>downwards</u>.

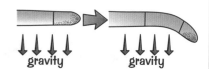

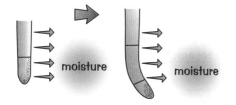

Roots grow towards water

1) An uneven amount of moisture either side of a root produces <u>more auxin</u> on the side with more <u>moisture</u>.
2) This <u>inhibits</u> growth on that side, causing the root to grow in that direction, <u>towards the moisture</u>.

Plant Hormones Can be Extracted and Used by People

1) <u>Seedless fruits</u> can be made with <u>artificial hormones</u>:

> Fruit (with seeds in the middle) normally only grows on plants which have been <u>pollinated by insects</u>. If the plant <u>doesn't</u> get pollinated, the fruit and seeds <u>don't</u> grow. If <u>growth hormones</u> are applied to the <u>unpollinated flowers</u> of some types of plant, the <u>fruit will grow</u> but the <u>seeds won't</u>.

2) <u>Selective weedkillers</u> have been developed from <u>plant growth hormones</u>.
3) <u>Cuttings</u> (see p.41) can be grown using <u>rooting powder</u> (containing a <u>plant growth hormone</u>).
4) <u>Fruit</u> can be <u>ripened</u> on its way to the shops using a <u>ripening hormone</u>.
5) <u>Seeds</u> can be forced to <u>germinate</u> using a hormone called <u>gibberellin</u>.

A plant auxin to a bar — 'ouch'...

Plants grow in places where there are <u>good growing conditions</u> — like lots of nutrients, light and O_2.

Selective Breeding

Selective breeding is a way for humans to develop crops/herds with <u>useful characteristics</u>.

Selective Breeding *is Mating the Best Organisms to Get Good Offspring*

<u>Selective breeding</u> is when humans select the plants or animals that are going to breed and flourish, according to what <u>we</u> want from them. It's also called <u>artificial selection</u>.

This is the basic process involved in selective breeding:

1) From the existing stock, the organisms which have the <u>best characteristics</u> are selected.

2) They're <u>bred</u> with each other.

3) The <u>best</u> of the <u>offspring</u> are selected and <u>bred</u>.

4) This process is repeated over several generations to develop the <u>desired traits</u>.

Selective Breeding *is Very Useful in Farming*

Farmers can improve the quality of milk from cattle

1) Cows can be selectively bred to produce offspring with particular characteristics, e.g. a <u>high milk yield</u> or <u>milk high in nutrients</u> (such as calcium or protein).

2) Cows are usually impregnated using <u>artificial insemination</u>. Semen from a <u>bull</u> which has <u>good characteristics</u> (or whose mother had good characteristics — bulls obviously don't produce milk) is used to artificially inseminate a <u>large number of cows</u>.

3) A typical dairy cow now produces between <u>5000 and 6000 litres</u> of milk a year. This is much more than dairy cows produced a hundred years ago. This is partly because of selective breeding and partly due to <u>intensive farming methods</u> (e.g. giving cows a special diet).

Farmers can increase the number of offspring in sheep

Farmers can selectively breed <u>sheep</u> to <u>increase</u> the number of <u>lambs born</u>. Female sheep (ewes) who produce large numbers of offspring are bred with rams whose mothers had large numbers of offspring. The <u>characteristic</u> of having large numbers of offspring is <u>passed on</u> to the next generation.

Farmers can increase the yield from dwarf wheat

1) Selective breeding can be used to combine <u>two different desirable characteristics</u>.

2) <u>Tall wheat plants</u> have a good grain yield but are easily damaged by wind and rain. <u>Dwarf wheat plants</u> can resist wind and rain but have a lower grain yield.

3) These two types of wheat plant were <u>cross-bred</u>, and the best resulting wheat plants were cross-bred again. This resulted in a <u>new variety</u> of wheat <u>combining the good characteristics</u> — dwarf wheat plants which could <u>resist bad weather</u> and had a <u>high grain yield</u>.

There are Disadvantages to Selective Breeding

1) Only some of the original population is bred from — so there's <u>less variety</u> in the <u>gene pool</u> of the organisms. All the organisms in a crop/herd will be <u>closely related</u> and have <u>similar characteristics</u> — including their <u>level of disease-resistance</u>. Some diseases might be able to <u>wipe out</u> the whole lot.

2) Some of the characteristics encouraged by selective breeding are <u>beneficial for humans</u>, but <u>not</u> for the <u>organisms</u> themselves. E.g. selective breeding to <u>increase milk yields</u> means cows produce more milk than they would need to feed a calf. They often suffer from <u>mastitis</u> (inflammation of the udders).

I use the same genes all the time too — they flatter my hips...

Selective breeding's <u>not</u> a <u>new</u> thing. People have been doing it for yonks. That's how we ended up with something like a <u>poodle</u> from a <u>wolf</u>. Somebody thought 'I really like this small, woolly, yappy wolf — I'll breed it with this other one'. And after <u>thousands of generations</u>, we got poodles. Hurrah.

Adult Cloning

Ah, <u>Dolly the sheep</u>. It seems a long time ago now, but she was the <u>first mammal cloned</u> from an <u>adult cell</u>. She was born in <u>1996</u>, the <u>only success</u> of <u>277 attempts</u> by the team who created her.

Cloning an Adult is Done by Transplanting a Cell Nucleus

The <u>first mammal</u> to be successfully cloned from an <u>adult cell</u> was a sheep called "Dolly".
This is the method that was used to produce Dolly:

1) The <u>nucleus</u> of a sheep's <u>egg cell</u> was removed — this left the egg cell without any <u>genetic information</u>.
2) Another nucleus was <u>inserted</u> in its place. This was a <u>diploid</u> nucleus from an udder cell of a <u>different sheep</u> (the one being cloned) and had all its <u>genetic information</u>.
3) The cell was <u>stimulated</u> so that it started <u>dividing by mitosis</u>, as if it was a normal <u>fertilised egg</u>.
4) The dividing cell was <u>implanted</u> into the <u>uterus</u> of another sheep to develop until it was ready to be born.
5) The result was <u>Dolly</u>, a clone of the sheep from which the <u>udder cell</u> came.

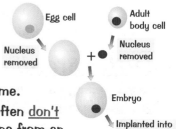

There are <u>risks</u> with cloning. Embryos formed by cloning from adult cells often <u>don't develop normally</u>. There had been many <u>failed attempts</u> at producing a clone from an <u>adult</u> before the success with Dolly.

There are Both Benefits and Risks Involved in Cloning

There are many possible <u>benefits</u> of cloning:

1) Animals that can produce <u>medicines</u> in their <u>milk</u> could be cloned. Researchers have managed to transfer <u>human genes</u> that produce <u>useful proteins</u> into <u>sheep</u> and <u>cows</u>, so that they can produce, for example, the blood clotting agent <u>factor VIII</u> used for treating <u>haemophilia</u>. With cloning, you only need to transfer the genes <u>once</u>, and then you could <u>clone</u> the animal as many times as you liked.
2) Animals (probably pigs) that have organs suitable for <u>organ transplantation</u> into humans (<u>xenotransplantation</u>) could be developed by <u>genetic engineering</u> and then <u>cloned</u> in the same way.
3) The <u>study</u> of animal clones and cloned cells could lead to <u>greater understanding</u> of the <u>development</u> of the <u>embryo</u> and of <u>ageing</u> and <u>age-related disorders</u>.
4) Cloning could be used to help preserve <u>endangered species</u>.

But there are <u>risks</u> too:

1) There is some evidence that cloned animals might <u>not</u> be as <u>healthy</u> as normal ones.
2) Cloning is a <u>new</u> science and it might have consequences that we're <u>not yet aware of</u>.
3) People are worried that <u>humans</u> might be produced by cloning if research continues.

Cloning Humans is a Possibility — with a Lot of Ethical Issues

As the technology used to clone mammals <u>improves</u>, it becomes more and more likely that <u>humans</u> could one day be <u>cloned</u> as well. However, there are still enormous <u>difficulties</u> to be overcome, and it might well involve women willing to <u>donate</u> hundreds of <u>eggs</u>. There would have to be lots of <u>surrogate pregnancies</u>, probably with <u>high rates</u> of <u>miscarriage</u> and <u>stillbirth</u>. The problems scientists have had with other mammals (see below) have shown that the human clones produced could well be <u>unhealthy</u> and <u>die prematurely</u>. There are also worries that if we clone humans we will be '<u>playing God</u>', and meddling with things we <u>don't fully understand</u>. Even if a healthy clone were produced, it might be <u>psychologically damaged</u> by the knowledge that it's just a clone of <u>another</u> human being.

A whole lamb from a single cell? Pull the udder one...

Since Dolly, scientists have successfully cloned <u>all kinds</u> of mammals including goats, cows, mice, pigs, cats, rabbits, horses and dogs. Many of these clones suffered <u>health problems</u> and <u>died young</u> — Dolly <u>seemed</u> normal, but died aged just <u>six</u> (when the breed has a life expectancy of 11-12).

Revision Summary for Section Eight

And that's another section finished. Award yourself a gold star, relax, get a cup of tea, and take a leisurely glance through these beautiful revision summary questions. Once you've glanced through them, you'll have to answer them. And then you'll have to check your answers and go back and revise any bits you got wrong. And then do the questions again. In fact, it's not really a matter of relaxing at all. More a matter of knuckling down to lots of hard work. Oops. Sorry.

1) Give three ways that the growth of an organism can be measured.
2) Describe two differences in the way plant cells and animal cells grow and develop.
3) Give an example of an animal which can regenerate.
4) Why is it illegal for athletes to use growth factors to enhance their performance?
5) What is mitosis used for in the human body? Describe the four steps in mitosis.
6) What is the Hayflick limit? What two types of cell don't have a Hayflick limit?
7) Where does meiosis take place in the human body?
8) What type of cell division does a fertilised egg use to grow into a new organism?
9) Give three ways that sperm cells are adapted to their function.
10) Why is the legal time limit for terminating a foetus set at 24 weeks?
11) During in vitro fertilisation, it is possible to screen embryos for various genetic disorders before they're implanted into the mother. Only the "good" embryos would be chosen for implantation. Summarise the main arguments for and against embryonic screening.
12) What is meant by the 'differentiation' of cells?
13) How are the stem cells in an embryo different from the stem cells in an adult?
14) Give three examples of how embryonic stem cells could be used to cure diseases.
15) There are concerns about the ethics of stem cell research. Give one argument in favour of stem cell research and one argument against stem cell research.
16) Explain how auxins cause plant shoots to grow towards light.
17) Explain how auxins cause plant roots to grow towards water.
18) Give three ways that plant hormones are used commercially.
19) What is selective breeding?
20) Suggest three features that you might selectively breed for in a dairy cow.
21) Give three examples of the use of selective breeding in farming.
22) Describe two disadvantages of selective breeding.
23) Describe the process of cloning an animal from an adult cell (e.g. cloning a sheep).
24) Describe three risks associated with trying to clone animals.

Photosynthesis

Plants can make their own food — it's ace. Here's how...

Photosynthesis Produces Glucose Using Sunlight

1) Photosynthesis is the process that produces 'food' in plants. The 'food' it produces is glucose.
2) Photosynthesis happens in the leaves of all green plants — this is largely what the leaves are for.
3) Photosynthesis happens inside the chloroplasts, which are found in leaf cells and in other green parts of a plant. Chloroplasts contain a substance called chlorophyll, which absorbs sunlight and uses its energy to convert carbon dioxide and water into glucose. Oxygen is also produced.

$$\text{carbon dioxide} + \text{water} \xrightarrow[\text{chlorophyll}]{\text{SUNLIGHT}} \text{glucose} + \text{oxygen}$$

Four Things are Needed for Photosynthesis to Happen:

1) Light

Sunlight beating down on the leaf provides the energy for the process.

2) Chlorophyll

This is the green substance which is found in chloroplasts and which makes leaves look green.
Chlorophyll absorbs the energy in sunlight and uses it to combine CO_2 and water to make glucose. Oxygen is just a by-product of this reaction.

3) Carbon dioxide

CO_2 diffuses into the leaf from the air around.

4) Water

Water comes up from the soil, up the roots and stem, and into the leaf via the veins.

Plants Use the Glucose for Six Different Things:

1) For respiration — this releases energy that enables them to convert the rest of the glucose into various other useful substances which they use to build new cells and grow.
2) Making fruits — glucose, along with another sugar called fructose, is turned into sucrose for storing in fruits. Fruits deliberately taste nice so that animals will eat them and spread the seeds all over the place in their poo.
3) Making cell walls — glucose in converted into cellulose for making cell walls.
4) Making proteins — glucose is combined with nitrates to make amino acids which are then made into proteins.
5) Stored in seeds — glucose is turned into lipids (fats and oils) for storing in seeds.
6) Stored as starch — glucose is turned into starch and stored in roots, stems and leaves, ready for use when photosynthesis isn't happening, like in winter. Starch is insoluble, which makes it much better for storing because it doesn't bloat the storage cells by osmosis like glucose would.

I'm working on sunshine... woah o...

Plants are pretty crucial in ensuring the flow of energy through nature. They are able to use the Sun's energy to make glucose — the energy source which humans and animals need for respiration (see p.60). Make sure you know the photosynthesis equation inside out — it's important later in the section too.

Rate of Photosynthesis

A plant's rate of photosynthesis is affected by the amount of <u>light</u>, the amount of <u>CO_2</u>, and the <u>temperature</u> of its surroundings. Photosynthesis slows down or stops if the conditions aren't right.

The Limiting Factor Depends on the Conditions

1) A limiting factor is something which <u>stops photosynthesis from happening any faster</u>. The amount of light, amount of CO_2 and the temperature can all be the limiting factor.

2) The limiting factor depends on the <u>environmental conditions</u>. E.g. in <u>winter</u> cold temperatures might be the limiting factor, at <u>night</u> light is likely to be the limiting factor.

There are Three Important Graphs for Rate of Photosynthesis

1) Not Enough LIGHT Slows Down the Rate of Photosynthesis

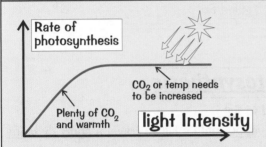

<u>Chlorophyll</u> uses <u>light energy</u> to perform photosynthesis. It can only do it as quickly as the light energy is arriving.

1) If the <u>light level</u> is raised, the rate of photosynthesis will <u>increase steadily</u>, but only up to a <u>certain point</u>.

2) Beyond that, it won't make any <u>difference</u> because then it'll be either the <u>temperature</u> or the <u>CO_2</u> level which is now the limiting factor.

2) Too Little CARBON DIOXIDE Also Slows It Down

<u>CO_2</u> is one of the <u>raw materials</u> needed for photosynthesis — only <u>0.04%</u> of the air is CO_2, so it's <u>pretty scarce</u> as far as plants are concerned.

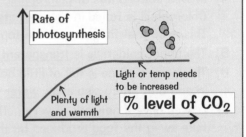

1) As with light intensity, increasing the amount of CO_2 will only <u>increase</u> the rate of photosynthesis up to a point. After this the graph <u>flattens out</u>, showing that CO_2 is no longer the limiting factor.

2) As long as <u>light</u> and <u>CO_2</u> are in plentiful supply then the factor limiting photosynthesis must be <u>temperature</u>.

3) The TEMPERATURE Has to be Just Right

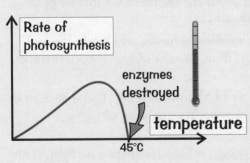

Temperature affects the rate of photosynthesis — because it affects the <u>enzymes</u> involved.

1) As the <u>temperature increases</u>, so does the <u>rate</u> of photosynthesis — up to a point.

2) If the temperature is <u>too high</u> (over about 45 °C), the plant's <u>enzymes</u> will be <u>denatured</u> (destroyed), so the rate of photosynthesis rapidly decreases.

3) <u>Usually</u> though, if the temperature is the <u>limiting factor</u> it's because it's too low, and things need <u>warming up a bit</u>.

No, no... no, no, no, no... no, no, no, no... no, no, there's no limit...

You can create the best conditions for photosynthesis in a greenhouse. Farmers use heaters and artificial lights and they can also increase the level of CO_2 using paraffin burners. By keeping plants in a greenhouse, they're also keeping out pests and diseases. The plants will grow much more quickly.

Leaf Structure

Now's a good time to flick back to p.56 and make sure that you thoroughly know about diffusion.

Leaves are Designed for Making Food by Photosynthesis

The whole structure of leaves is geared towards that.
You need to know all the different parts of a
typical leaf shown on the diagram:

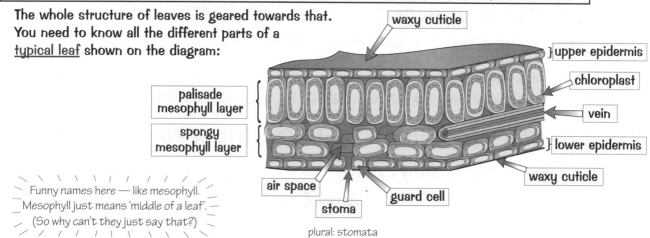

Funny names here — like mesophyll.
Mesophyll just means 'middle of a leaf'.
(So why can't they just say that?)

plural: stomata

Leaves are Adapted for Efficient Photosynthesis

1) Leaves are broad, so there's a large surface area exposed to light.

2) They're also thin, which means carbon dioxide and water vapour only have to travel a short distance to reach the photosynthesising cells where they're needed.

3) There are air spaces in the spongy mesophyll layer. This lets gases like carbon dioxide and oxygen move easily between cells. This also means there's a large surface area for gas exchange.

4) Leaves contain lots of chlorophyll, which is the pigment that absorbs light energy for photosynthesis. Chlorophyll is found in chloroplasts, and most of the chloroplasts are found in the palisade layer. This is so that they're near the top of the leaf where they can get the most light.

5) The upper epidermis is transparent so that light can pass through it to the palisade layer.

6) The lower surface is full of little holes called stomata. They're there to let gases like CO_2 and O_2 in and out. They also allow water to escape — which is known as transpiration.

7) Leaves have a network of veins. These deliver water and other nutrients to every part of the leaf and take away the food produced by the leaf. They also help to support the leaf structure.

Plants Exchange Gases by Diffusion

When plants photosynthesise they use up CO_2 from the atmosphere and produce O_2 as a product. Plants also respire where they use up O_2 and produce CO_2 as a product. So there are lots of gases moving to and fro in plants, and this movement happens by diffusion.

1) When the plant is photosynthesising it uses up lots of CO_2, so there's hardly any inside the leaf. Luckily this makes more CO_2 move into the leaf by diffusion (from an area of higher concentration to an area of lower concentration).

2) At the same time lots of O_2 is being made as a waste product of photosynthesis. Some is used in respiration, and the rest diffuses out through the stomata (moving from an area of higher concentration to an area of lower concentration).

3) At night it's a different story — there's no photosynthesis going on because there's no light. Lots of carbon dioxide is made in respiration and lots of oxygen is used up. There's a lot of CO_2 in the leaf and not a lot of O_2, so now it's mainly carbon dioxide diffusing out and oxygen diffusing in.

If you don't do much revision, it's time to turn over a new leaf...

Scientists know all this stuff because they've looked and seen the structure of leaves and the cells inside them. Not with the naked eye, of course — they used microscopes. So they're not just making it all up.

Transpiration

If you don't water a house plant for a few days it starts to go <u>all droopy</u>. Then it <u>dies</u>, and the people from the Society for the Protection of Plants come round and have you <u>arrested</u>. Plants need water.

Transpiration *is the Loss of Water from the Plant*

1) Transpiration is caused by the <u>evaporation</u> and <u>diffusion</u> (see page 56) of water from inside the leaves.

2) This creates a slight <u>shortage</u> of water in the leaf, and so more water is drawn up from the rest of the plant through the <u>xylem vessels</u> (see next page) to replace it.

3) This in turn means more water is drawn up from the <u>roots</u>, and so there's a constant <u>transpiration stream</u> of water through the plant.

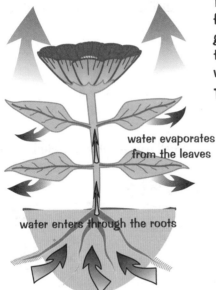

water evaporates from the leaves

water enters through the roots

Transpiration is just a <u>side-effect</u> of the way leaves are adapted for <u>photosynthesis</u>. They have to have <u>stomata</u> in them so that gases can be exchanged easily (see previous page). Because there's more water <u>inside</u> the plant than in the <u>air outside</u>, the water escapes from the leaves through the stomata.

The transpiration stream does have some <u>benefits</u> for the plants:

1) The constant stream of water from the ground helps to keep the plant <u>cool</u>.

2) It provides the plant with a constant supply of water for <u>photosynthesis</u>.

3) The water creates <u>turgor pressure</u> in the plant cells, which helps support the plant and stops it wilting (see next page).

4) <u>Minerals</u> needed by the plant (see page 83) can be brought in from the soil along with the water.

Transpiration Rate *is Affected by Four Main Things*

1) <u>LIGHT INTENSITY</u> — the <u>brighter</u> the light, the <u>greater</u> the transpiration rate.
<u>Stomata</u> begin to <u>close</u> as it gets darker. Photosynthesis can't happen in the dark, so they don't need to be open to let CO_2 in. When the stomata are closed, very little water can escape.

2) <u>TEMPERATURE</u> — the <u>warmer</u> it is, the <u>faster</u> transpiration happens.
When it's warm the water particles have <u>more energy</u> to evaporate and diffuse out of the stomata.

3) <u>AIR MOVEMENT</u> — if there's <u>lots</u> of air movement (wind) around a leaf, transpiration happens <u>faster</u>.
If the air around a leaf is very still, the water vapour just <u>surrounds the leaf</u> and doesn't move away. This means there's a <u>high concentration</u> of water particles outside the leaf as well as inside it, so <u>diffusion</u> doesn't happen as quickly. If it's windy, the water vapour is <u>swept away</u>, maintaining a <u>low concentration</u> of water in the air outside the leaf. Diffusion then happens quickly, from an area of high concentration to an area of low concentration.

4) <u>AIR HUMIDITY</u> — if the air around the leaf is very <u>dry</u>, transpiration happens more <u>quickly</u>.
This is like what happens with air movement. If the air is <u>humid</u> there's a lot of water in it already, so there's not much of a <u>difference</u> between the inside and the outside of the leaf. Diffusion happens <u>fastest</u> if there's a <u>really high concentration</u> in one place, and a <u>really low concentration</u> in the other.

Transpiration — the plant version of perspiration...

One good way to remember those <u>four factors</u> that affect the rate of transpiration is to think about drying washing. Then you'll realise there are far more boring things you could be doing than revision, and you'll try harder. No, only joking — it's the same stuff: <u>sunny</u>, <u>warm</u>, <u>windy</u> and <u>dry</u>.

Water Flow in Plants

Plants Need to Balance Water Loss with Water Uptake

Transpiration can help plants in some ways (see previous page), but if it hasn't rained for a while and you're <u>short of water</u> it's not a good idea to have it rushing out of your leaves. So plants have <u>adaptations</u> to help <u>reduce water loss</u> from their leaves.

1) Leaves usually have a <u>waxy waterproof cuticle</u> covering the <u>upper epidermis</u>.

2) Most <u>stomata</u> are found on the <u>lower surface</u> of a leaf where it's <u>darker</u> and <u>cooler</u>. This helps slow down <u>diffusion</u> of water out of the leaf.

3) The <u>bigger</u> the stomata and the <u>more</u> stomata a leaf has, the more <u>water</u> the plant will <u>lose</u>. Plants in <u>hot climates</u> really need to conserve water, so they have <u>fewer</u> and <u>smaller</u> stomata on the underside of the leaf and <u>no</u> stomata on the upper epidermis.

Turgor Pressure Supports Plant Tissues

Normal Cell   Turgid Cell

1) When a plant is well watered, all its cells will draw water in by <u>osmosis</u> and become plump and swollen. When the cells are like this, they're said to be <u>turgid</u>.

2) The contents of the cell push against the cell wall — this is called <u>turgor pressure</u>. Turgor pressure helps <u>support</u> the plant tissues.

3) If there's no water in the soil, a plant starts to <u>wilt</u> (droop). This is because the cells start to lose water and so <u>lose</u> their turgor pressure. The cells are then said to be <u>flaccid</u>.

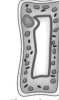

4) If the plant's really short of water, the <u>cytoplasm</u> inside its cells starts to <u>shrink</u> and the membrane <u>pulls away</u> from the cell wall. The cell is now said to be <u>plasmolysed</u>. The plant doesn't totally lose its shape though, because the <u>inelastic cell wall</u> keeps things in position. It just droops a bit.

Flaccid Cell   Plasmolysed Cell

Plants Have Xylem and Phloem

Plants have <u>two</u> separate types of vessel — <u>xylem</u> and <u>phloem</u> — for transporting stuff around. <u>Both</u> types of vessel go to <u>every part</u> of the plant, but they are totally <u>separate</u>.

Water and minerals

Xylem tubes take water up:

1) Made of <u>dead cells</u> joined end to end with <u>no</u> end walls between them and a hole (<u>lumen</u>) down the middle.

2) The thick side walls are strong and stiff, which gives the plant <u>support</u>.

3) They carry <u>water</u> and <u>minerals</u> from the <u>roots</u> up the shoot to the leaves in the <u>transpiration stream</u>.

Phloem tubes transport food:

1) Made of columns of living cells with <u>perforated end-plates</u> to allow stuff to flow through.

2) They transport <u>food substances</u> (mainly <u>sugars</u>) made in the leaves to growing and storage tissues, in <u>both directions</u>.

3) This movement of food substances around the plant is known as <u>translocation</u>.

Food (mainly sugars)

Don't let revision stress you out — just go with the phloem...

You probably did that really dull experiment at school where you stick a piece of celery in a beaker of water with red food colouring in it. Then you stare at it for half an hour, and once the time is up, hey presto, the red has reached the top of the celery. That's because it travelled there in the xylem.

Minerals for Healthy Growth

Plants are important in <u>food chains</u> and <u>nutrient cycles</u> because they can take <u>minerals</u> from the soil and <u>energy</u> from the Sun and turn it into food. And then, after all that hard work, we eat them.

Plants Need Three Main Minerals

Plants need certain <u>elements</u> so they can produce important compounds. They get these elements from <u>minerals</u> in the <u>soil</u>. If there aren't enough of these minerals in the soil, plants suffer <u>deficiency symptoms</u>.

1) Nitrates

Contain nitrogen for making <u>amino acids</u> and <u>proteins</u>. These are needed for <u>cell growth</u>. If a plant can't get enough nitrates it will be <u>stunted</u> and will have <u>yellow older leaves</u>.

2) Phosphates

Contain phosphorus for making <u>DNA</u> and <u>cell membranes</u> and they're needed for <u>respiration</u> and <u>growth</u>. Plants without enough phosphate have <u>poor root growth</u> and <u>purple older leaves</u>.

3) Potassium

To help the <u>enzymes</u> needed for <u>photosynthesis</u> and <u>respiration</u>. If there's not enough potassium in the soil, plants have <u>poor flower and fruit growth</u> and <u>discoloured leaves</u>.

Magnesium is Also Needed in Small Amounts

The three main minerals are needed in fairly <u>large amounts</u>, but there are other elements which are needed in much <u>smaller</u> amounts. <u>Magnesium</u> is one of the most significant as it's required for making <u>chlorophyll</u> (needed for <u>photosynthesis</u>). Plants without enough magnesium have <u>yellow leaves</u>.

Root Hairs Take In Minerals and Water

1) The cells on plant roots grow into long '<u>hairs</u>' which stick out into the soil.

2) Each branch of a root will be covered in <u>millions</u> of these microscopic hairs.

3) This gives the plant a <u>big surface area</u> for absorbing <u>minerals</u> and <u>water</u> from the soil.

Minerals are taken in by active transport

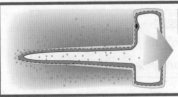

1) The <u>concentration</u> of minerals in the <u>soil</u> is usually pretty <u>low</u>. It's normally <u>higher</u> in the <u>root hair cell</u> than in the soil around it.

2) So normal diffusion <u>doesn't</u> explain how minerals are taken up into the root hair cell. They should go <u>the other way</u> if they followed the rules of diffusion.

3) The answer is that a different process called '<u>active transport</u>' is responsible.

4) Active transport uses <u>energy</u> from <u>respiration</u> to help the plant pull minerals into the root hair <u>against the concentration gradient</u>. This is essential for its growth.

Water is taken in by osmosis

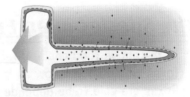

There's usually a <u>higher concentration</u> of water in the soil than there is inside the plant, so the water is drawn into the root hair cell by <u>osmosis</u>.

Nitrogen and phosphorus and potassium — oh my...

When a farmer or a gardener buys <u>fertiliser</u>, that's pretty much what he or she is buying — <u>nitrates</u>, <u>phosphates</u> and <u>potassium</u>. A fertiliser's <u>NPK label</u> tells you the relative proportions of nitrogen (N), phosphorus (P) and potassium (K) it contains, so you can choose the right one for your plants and soil.

Pyramids of Number and Biomass

A <u>trophic level</u> is a <u>feeding</u> level. It comes from the Greek word <u>trophe</u> meaning 'nourishment'. The amount of <u>energy</u>, <u>biomass</u> and usually the <u>number of organisms</u> all <u>decrease</u> as you move up a trophic level.

You Need to be Able to Construct Pyramids of Number

Luckily it's pretty easy — they'll give you all the information you need to do it in the exam. Here's an example:

<u>5000</u> dandelions... feed... <u>100</u> rabbits... which feed... <u>1</u> fox.

1) Each bar on a pyramid of numbers shows the <u>number of organisms</u> at that stage of the food chain.

2) So the '<u>dandelions</u>' bar on this pyramid would need to be <u>longer</u> than the '<u>rabbits</u>' bar, which in turn should be <u>longer</u> than the '<u>fox</u>' bar.

3) <u>Dandelions</u> go at the <u>bottom</u> because they're at the bottom of the food chain.

This gives a <u>typical pyramid of numbers</u>, where every time you go up a <u>trophic (feeding) level</u>, the number of organisms goes <u>down</u>. This is because it takes a <u>lot</u> of food from the level below to keep one animal alive.

1 fox
100 rabbits
5000 dandelions

There are cases where a number pyramid is <u>not a pyramid at all</u>. For example 1 fox may feed 500 fleas.

You'll Have to Construct Pyramids of Biomass Too

1) Each bar on a <u>pyramid of biomass</u> shows the <u>mass of living material</u> at that stage of the food chain — basically how much all the organisms at each level would '<u>weigh</u>' if you put them <u>all together</u>.

2) So the one fox above would have a <u>big biomass</u> and the <u>hundreds of fleas</u> would have a <u>very small biomass</u>. Biomass pyramids are practically <u>always the right shape</u>:

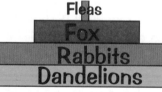

Fleas
Fox
Rabbits
Dandelions

The Species in an Environment are Interdependent

1) In the above food chain, if <u>rabbits</u> were wiped out, then the <u>foxes</u> would soon follow as they have <u>nothing to eat</u>. Foxes are said to be <u>dependent</u> on rabbits for survival.

2) Unfortunately life just isn't as simple as that. There are many different species within an environment, all <u>interdependent</u>. This means if one species changes, it <u>affects all the others</u>. For example, if lots of rabbits died, then:

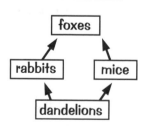

- There would be <u>less food</u> for the <u>foxes</u>, so their numbers might <u>decrease</u>.
- The number of <u>dandelions</u> might <u>increase</u>, because the rabbits wouldn't be eating them.
- The <u>mice</u> wouldn't be <u>competing</u> with the rabbits for food, so their numbers might <u>increase</u>.

Constructing pyramids is a breeze — just ask the Egyptians...

There are actually a couple of exceptions where pyramids of <u>biomass</u> aren't quite pyramid-shaped. It happens when the producer has a very short life but reproduces loads, like with plankton at certain times of year. But it's <u>rare</u>, and you <u>don't</u> need to know about it. Forget I ever mentioned it. Sorry.

Energy Transfer and Energy Flow

All That Energy Just Disappears Somehow...

1) Energy from the <u>Sun</u> is the source of energy for nearly <u>all</u> life on Earth.

2) <u>Plants</u> use a small percentage of the light energy from the Sun to make <u>food</u> during photosynthesis. This energy then works its way through the food web as animals eat the plants and each other.

3) Much of the <u>energy lost</u> at each stage is used for <u>staying alive</u>, i.e. in <u>respiration</u> (see page 60), which powers all life processes.

Material and energy are both lost at each stage of the food chain.

This explains why you get <u>biomass pyramids</u>. Most of the biomass is lost and so does <u>not</u> become biomass in the <u>next level up</u>.

HEAT LOSS

MATERIALS LOST IN ANIMAL'S WASTE

4) Most of this energy is eventually <u>lost</u> to the surroundings as <u>heat</u>. This is especially true for <u>mammals</u> and <u>birds</u>, whose bodies must be kept at a <u>constant temperature</u> which is normally higher than their surroundings.

5) <u>Material</u> and <u>energy</u> are also lost from the food chain in the <u>droppings</u> — you'll need to remember the posh word for producing droppings, which is <u>egestion</u>.

It also explains why you hardly ever get <u>food chains</u> with more than about <u>five trophic levels</u>. So much energy is <u>lost</u> at each stage that there's not enough left to support more organisms after four or five stages.

You Need to be Able to Interpret Data on Energy Flow

rosebush: 80 000 kJ   greenfly: 10 000 kJ   ladybird: 900 kJ   bird: 40 kJ

1) The numbers show the <u>amount of energy</u> available to the <u>next level</u>. So <u>80 000 kJ</u> is the amount of energy available to the <u>greenfly</u>, and <u>10 000 kJ</u> is the amount available to the <u>ladybird</u>.

2) You can work out how much energy has been <u>lost</u> at each level by taking away the energy that is available to the <u>next</u> level from the energy that was available from the <u>previous</u> level. Like this:

Energy <u>lost</u> at 1st trophic level as <u>heat</u> and in <u>egestion</u> = 80 000 kJ – 10 000 kJ = <u>70 000 kJ lost</u>.

3) You can also calculate the <u>efficiency of energy transfer</u> — this just means how good it is at passing on energy from one level to the next.

$$\text{efficiency} = \frac{\text{energy available to the next level}}{\text{energy that was available to the previous level}} \times 100$$

So at the 1st trophic level, <u>efficiency</u> of energy transfer = 10 000 kJ ÷ 80 000 kJ × 100
= <u>12.5% efficient</u>.

So when revising, put the fire on and don't take toilet breaks...

No, I'm being silly — go if you have to. We're talking in <u>general terms</u> about <u>whole food chains</u> here — you won't lose your concentration as a direct result of, erm, egestion.

Biomass and Fermentation

Energy Stored in Biomass Can be Used for Other Things

There are many different ways to release the energy stored in biomass — including eating it, feeding it to livestock, growing the seeds of plants and using it as a fuel.

For a given area of land, you can produce a lot more food for humans by growing crops than by grazing animals — only about 10% of the biomass eaten by beef cattle becomes useful meat for people to eat. It's important to get a balanced diet, though, which is difficult from crops only. It's also worth remembering that some land, like moorland or fellsides, isn't suitable for growing crops. In these places, animals like sheep and deer can be the best way to get food from the land.

As well as using biomass as food, you can use it as fuel. Learn these two examples of biofuels:

1) **Fast-growing trees** — people tend to think burning trees is a bad thing, but it's not as long as they're fast-growing and more are planted to replace them. There's no overall contribution to CO_2 emissions because the replacement trees are still removing carbon from the atmosphere.

2) **Fermenting biomass** — You can use microorganisms to make biogas from plant and animal waste in a simple fermenter (see below). The biogas can then be burned to release the energy.

Developing biofuels is a great idea, for these three important reasons:
- Unlike coal, oil and the like, biofuels are renewable — they're not going to run out one day.
- Using biofuels reduces air pollution — no acid rain gases are produced when wood and biogas burn.
- You can be energy self-reliant. Theoretically, you could supply your energy from household waste.

Mycoprotein is Grown in Fermenters

1) Mycoprotein is protein from a fungus, used to make meat substitutes for vegetarian meals, e.g. Quorn.
2) The fungus is grown in huge vessels called fermenters, using glucose syrup as food.
3) Microorganisms like fungi can grow very quickly... which is great if you're using them to make food.
4) They're also easy to look after. All that's needed is something to grow them in, food, oxygen, and the right temperature. So food can be produced whether the climate is hot or cold.
5) Microorganisms can use waste products from agriculture and industry as food.
6) This often makes using microorganisms cheaper than other methods.

A fermenter is a big container full of liquid 'culture medium' in which microorganisms can grow and reproduce. The fermenter needs to have the right conditions for the microorganisms to grow and produce their useful product.

1) Food (e.g. carbohydrate, mineral ions, nitrates) is provided in the liquid culture medium.

2) Air is piped in to supply oxygen (if needed).

3) The medium needs to be kept at the right temperature and pH for optimum growth of the microorganisms. Fermenters can be cooled with a water jacket which cold water is pumped through.

4) Sterile conditions are needed to prevent contamination.

5) A motorised stirrer keeps the microorganisms from sinking to the bottom.

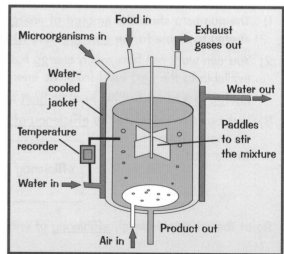

Be energy self-reliant — burn poo...

Microorganisms are really useful — as well making biogas, and producing mycoprotein, they're used to make antibiotics, AND we can genetically engineer them to make human proteins like insulin (see p.43).

Managing Food Production

There are four main ways to maximise food production: 1) <u>increase the energy transfer</u>, 2) <u>reduce disease</u>, 3) improve <u>feeding/growing conditions</u>, 4) <u>control predators</u>.

- If you <u>reduce</u> the number of <u>stages in the food chain</u>, you reduce the amount of energy lost. For a given area of land, you can produce <u>more food</u> for humans by growing <u>crops</u> than by grazing <u>animals</u>.
- Food production can also be made more efficient by <u>reducing the amount of energy</u> animals use, e.g. if you keep animals warm and still, they won't use as much energy and won't need to eat as much.

Fish Farms Reduce Energy Loss, Disease and Predators

Fish is an increasingly popular dish, but fish stocks are dwindling. '<u>Fish farms</u>' were set up to rear fish in a controlled way and increase their production. <u>Salmon farming</u> in Scotland is a good example:

1) The fish are kept in <u>cages</u> in a <u>sea loch</u>, to <u>stop them using as much energy</u> swimming about.

2) The cage also <u>protects</u> them from <u>predators</u> like birds and seals.

3) They're fed a <u>diet</u> of food pellets that's <u>carefully controlled</u> to <u>maximise</u> energy transfer and to avoid <u>pollution</u> to the loch.

4) Young fish are reared in <u>special tanks</u> to ensure as many survive as possible.

5) Fish kept in cages are more prone to <u>disease</u> and <u>parasites</u>. One pest is <u>sea lice</u>, which can be treated with <u>pesticides</u> which kill them.
To <u>avoid pollution</u> from chemical pesticides, <u>biological pest control</u> (see next page) can be used instead, e.g. a small fish called a <u>wrasse</u> eats the lice off the backs of the salmon.

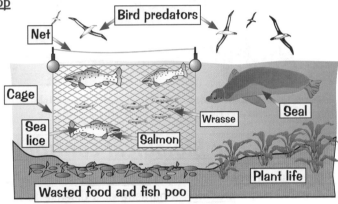

Food Production Involves Compromises and Conflict

Improving the efficiency of food production is useful — it means <u>cheaper food</u> for us, and <u>better standards of living</u> for farmers. But it all comes at a <u>cost</u>.

1) Some people think that forcing animals to live in unnatural and uncomfortable conditions is <u>cruel</u>. There's a growing demand for <u>organic meat</u>, from animals which have <u>not</u> been intensively farmed.

2) The <u>crowded</u> conditions on factory farms create a favourable environment for the <u>spread of diseases</u>, like avian flu and foot-and-mouth disease.

3) To try to <u>prevent disease</u>, animals are given <u>antibiotics</u>. When the animals are eaten these can enter humans. This allows <u>microbes</u> that infect humans to develop <u>immunity</u> to those antibiotics — so the antibiotics become <u>less effective</u> as <u>human</u> medicines.

4) The environment where the animals are kept needs to be <u>carefully controlled</u>. The animals need to be kept <u>warm</u> to reduce the energy they lose as heat. This often means using power from <u>fossil fuels</u> — which we wouldn't be using if the animals were grazing in their <u>natural</u> environment.

5) Our <u>fish stocks</u> are getting low. Yet a lot of fish goes on feeding animals that are <u>intensively farmed</u> — these animals wouldn't usually eat this source of <u>food</u>.

In an exam, you may be asked to give an account of the <u>positive</u> and <u>negative</u> aspects of food management. You will need to put <u>both sides</u>, whatever your <u>personal opinion</u> is. If you get given some <u>information</u> on a particular case, make sure you <u>use it</u> — they want to see that you've read it <u>carefully</u>.

Locked in a little cage with no sunlight — who'd work in a bank...

The world produces enough food to feed the Earth's population, but there are millions of undernourished people worldwide. The food is not equally distributed. Many people think that countries with food surpluses should give food to countries with food shortages (or sell them cheaply).

Pesticides and Biological Control

Biological control is growing more popular, as people get fed up with all the problems caused by pesticides.

Pesticides Disturb Food Chains

1) Pesticides are sprayed onto crops to kill the creatures that damage them, but unfortunately they also kill lots of harmless animals such as bees and beetles.

2) This can cause a shortage of food for animals further up the food chain.

3) Pesticides also tend to be toxic to creatures that aren't pests and there's a danger of the poison passing on through the food chain to other animals. There's even a risk that they could harm humans.

This is well illustrated by the case of otters which were almost wiped out over much of crop-dominated southern England by a pesticide called DDT in the early 1960s. The diagram shows the food chain which ends with the otter. DDT can't be excreted, so it accumulates along the food chain and the otter ends up with a lot of the DDT collected by the other animals.

③ Each little tiny animal eats lots of small plants
⑤ Each eel eats lots of small fish

① Insecticide seeps into the river
② Small water plants take up a little insecticide
④ Each small fish eats lots of tiny animals
⑥ Each otter eats lots of eels

You Can Use Biological Control Instead of Pesticides

Biological control means using living things instead of chemicals to control a pest.
You could use a predator, a parasite or a disease to kill the pest. For example:

1) Aphids are a pest because they eat roses and vegetables. Ladybirds are aphid predators, so people release them into their fields and gardens to keep aphid numbers down.

2) Certain types of wasps and flies produce larvae which develop on (or in, yuck) a host insect. This eventually kills the insect host. Lots of insect pests have parasites like this.

3) Myxomatosis is a disease which kills rabbits. The myxoma virus was released in Australia as a biological control when the rabbit population there grew out of control and ruined crops.

You need to be able to explain the advantages and disadvantages of biological control:

ADVANTAGES:
- The predator, parasite or disease usually only affects the pest animal. You don't kill all the harmless and helpful creatures as well like you often do with a pesticide.
- No chemicals are used, so there's less pollution, disruption of food chains and risk to people eating the food.

DISADVANTAGES:
- It's slower than pesticides — you have to wait for your control organism to build up its numbers.
- Biological control won't kill all the pests, and it usually only kills one type of pest.
- It takes more management and planning, and workers might need training or educating.
- Control organisms can drive out native species, or become pests in their own right.

Remember that removing an organism from a food web, whether you use biological control or pesticides, can affect all the other organisms too. For example, if you remove a pest insect, you're removing a source of food from all the organisms that normally eat it. These might die out, and another insect that they normally feed on could breed out of control and become a pest instead. You have to be very careful.

Don't get bugged by biological pest control...

In the exam you might be asked to interpret data related to biological control, e.g. tables showing the population sizes of pest species when using biological control and when using pesticides. Or they might give you a food web and ask you to predict the effect of removing different organisms.

Alternatives to Intensive Farming

Intensive farming methods are still used, a lot. But people are also using other methods more and more.

Hydroponics is Where Plants are Grown Without Soil

Most commercially grown <u>tomatoes</u> and <u>cucumbers</u> are grown in <u>nutrient solutions</u> (water and fertilisers) instead of in soil — this is called <u>hydroponics</u>.

There are <u>advantages</u> and <u>disadvantages</u> of using hydroponics instead of growing crops in soil:

ADVANTAGES	DISADVANTAGES
Takes up less space so less land required	It can be expensive to set up and run
No soil preparation or weeding needed	Need to use specially formulated soluble nutrients
Can still grow plants even in areas with poor soil	Growers need to be skilled and properly trained
Many pest species live in soil, so it avoids these	There's no soil to anchor the roots so plants need support
Mineral levels can be controlled more accurately	

Organic Farming is Still Perfectly Viable

Modern intensive farming produces lots of <u>food</u> and we all appreciate it on the supermarket shelves. But traditional <u>organic farming</u> methods do still work (amazingly!), and they have their <u>benefits</u> too. You need to know about these organic farming <u>techniques</u>:

1) Use of <u>organic fertilisers</u> (i.e. animal manure and compost). This <u>recycles</u> the nutrients left in plant and animal waste. It <u>doesn't work as well</u> as artificial fertilisers, but it is better for the <u>environment</u>.

2) <u>Crop rotation</u> — growing a cycle of <u>different crops</u> in a field each year. This stops the <u>pests</u> and <u>diseases</u> of one crop building up, and helps prevent <u>nutrients</u> running out (as each crop has slightly <u>different needs</u>). Most crop rotations include a <u>legume plant</u> like peas or beans, as they help put <u>nitrates</u> back in the soil (see page 83).

3) <u>Weeding</u> — this means <u>physically removing</u> the weeds, rather than just spraying them with a <u>herbicide</u>. Obviously it's a lot more <u>labour-intensive</u>, but there are no nasty <u>chemicals</u> involved.

4) <u>Varying seed planting times</u> — sowing seeds later or earlier in the season can <u>avoid</u> the <u>major pests</u> for that crop. This means the farmer <u>won't</u> need to use <u>pesticides</u>.

5) <u>Biological control</u> — this is covered on the previous page.

You also need to be able to discuss the <u>advantages</u> and <u>disadvantages</u> of organic farming. Always try to give a <u>balanced</u> point of view, unless you're specifically asked to argue one way or another. You can include your <u>own opinion</u> in a conclusion at the end. Here are a few points you could mention:

1) Organic farming takes up <u>more space</u> than intensive farming — so more land has to be <u>farmland</u>, rather than being set aside for wildlife or for other uses.

2) It's more <u>labour-intensive</u>. This provides <u>more jobs</u>, but it also makes the food more <u>expensive</u>.

3) You can't grow <u>as much</u> food. But on the other hand, Europe <u>over-produces</u> food these days anyway.

4) Organic farming uses fewer <u>chemicals</u>, so there's less risk of toxic chemicals remaining on food.

5) It's better for the <u>environment</u>. There's less chance of <u>polluting rivers</u> with fertiliser. Organic farmers also avoid using <u>pesticides</u>, so don't disrupt food chains and harm wildlife.

6) For a farm to be classed as organic, it will usually have to follow guidelines on the <u>ethical treatment of animals</u>. This means <u>no</u> battery farming.

Plants without soil? It's not like when I was a lad...

You can't just learn about the <u>methods</u> used in different types of farming — you have to think about their <u>impact</u> too. That means weighing up the advantages and disadvantages and being able to discuss them.

Recycling Nutrients

Carbon is constantly moving between the <u>atmosphere</u>, the <u>soil</u> and <u>living things</u> in the carbon cycle.

Nutrients are Constantly Recycled

1) <u>Living things</u> are made of materials they take from the world around them.

2) <u>Plants</u> take elements like <u>carbon</u>, <u>oxygen</u>, <u>hydrogen</u> and <u>nitrogen</u> from the <u>soil</u> or the <u>air</u>. They turn these elements into the <u>complex compounds</u> (carbohydrates, proteins and fats) that make up living organisms, and these then pass through the <u>food chain</u>.

3) These elements are <u>returned</u> to the environment in <u>waste products</u> produced by the organisms, or when the organisms <u>die</u>. The materials decay because they're <u>broken down</u> (digested) by <u>microorganisms</u> — that's how the elements get put back into the <u>soil</u>.

4) Microorganisms work best in <u>warm</u>, <u>moist</u> conditions. Some also need <u>oxygen</u>.

5) All the important <u>elements</u> are thus <u>recycled</u> — they return to the soil, ready to be <u>used</u> by new <u>plants</u> and put back into the <u>food chain</u> again.

<u>Detritivores</u> and <u>saprophytes</u> are both types of organism that are important in <u>decay</u>. They're grouped into those two types depending on <u>how they feed</u>.

1) <u>Detritivores</u>, e.g. earthworms, maggots and woodlice, feed on dead and decaying material (<u>detritus</u>). As these detritivores feed on the decaying material, they break it up into <u>smaller bits</u>. This gives a <u>bigger surface area</u> for smaller decomposers to work on and so <u>speeds up</u> decay.

2) <u>Saprophytes</u> feed on decaying material by <u>extracellular digestion</u>, i.e. they feed by <u>secreting digestive enzymes</u> onto the material outside of their cells. The enzymes <u>break down</u> the material into smaller bits which can then be <u>absorbed</u> by the saprophyte. Most saprophytes are <u>bacteria</u> and <u>fungi</u>.

The Carbon Cycle Shows How Carbon is Recycled

<u>Carbon</u> is an important element in the materials that living things are made from. It's constantly <u>recycled</u>:

This diagram isn't half as bad as it looks. <u>Learn</u> these important points:

1) There's only <u>one arrow</u> going <u>down</u>. The whole thing is 'powered' by <u>photosynthesis</u>. Green <u>plants</u> use the carbon from CO_2 in the air to make <u>carbohydrates</u>, <u>fats</u> and <u>proteins</u>.

2) Both plant and animal <u>respiration</u> while the organisms are alive <u>releases CO_2</u> back into the <u>air</u>.

3) Dead plants and animals can go <u>three ways</u>: be <u>eaten</u>, be <u>decayed</u> by <u>microorganisms</u> or be turned into <u>useful products</u> by humans.

4) <u>Eating</u> transfers some of the fats, proteins and carbohydrates to <u>new</u> fats, proteins and carbohydrates in the animals doing the eating.

5) Plant and animal products either <u>decay</u> or are <u>burned</u> (combustion) and <u>CO_2 is released</u>.

Come on out, it's only a little carbon cycle, it can't hurt you...

Carbon is a very <u>important element</u> for living things — it's the basis for all the organic molecules.

Recycling Nutrients

Nitrogen, just like carbon, is constantly being <u>recycled</u>. So the nitrogen in your proteins might once have been in the <u>air</u>. And before that it might have been in a <u>plant</u>. Or even in some <u>horse wee</u>. Nice.

Nitrogen is Also Recycled in the Nitrogen Cycle

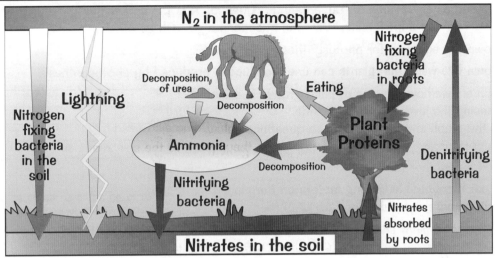

1) The <u>atmosphere</u> contains about <u>78% nitrogen gas</u>, N_2. This is <u>very unreactive</u> and so it can't be used <u>directly</u> by plants or animals.

2) <u>Nitrogen</u> is <u>needed</u> for making <u>proteins</u> for growth, so living organisms have to get it somehow.

3) Plants get their nitrogen from the <u>soil</u>, so nitrogen in the air has to be turned into <u>nitrogen compounds</u> before plants can use it. <u>Animals</u> can only get <u>proteins</u> by eating plants (or each other).

4) <u>Decomposers</u> break down <u>proteins</u> in rotting plants and animals, and <u>urea</u> in animal waste, into <u>ammonia</u>. So the nitrogen in these organisms is <u>recycled</u>.

5) <u>Nitrogen fixation</u> isn't an obsession with nitrogen — it's the process of turning <u>N_2 from the air</u> into <u>nitrogen compounds</u> in the soil which <u>plants can use</u>. There are <u>two main ways</u> that this happens:
   a) <u>Lightning</u> — there's so much <u>energy</u> in a bolt of lightning that it's enough to make nitrogen <u>react with oxygen</u> in the air to give nitrates.
   b) <u>Nitrogen-fixing bacteria</u> in roots and soil (see below).

6) There are <u>four</u> different types of <u>bacteria</u> involved in the nitrogen cycle:
   a) <u>DECOMPOSERS</u> — decompose <u>proteins</u> and <u>urea</u> and turn them into <u>ammonia</u>.
   b) <u>NITRIFYING BACTERIA</u> — turn <u>ammonia</u> in decaying matter into <u>nitrates</u>.
   c) <u>NITROGEN-FIXING BACTERIA</u> — turn <u>atmospheric N_2</u> into <u>nitrogen compounds</u> that plants can use.
   d) <u>DENITRIFYING BACTERIA</u> — turn <u>nitrates</u> back into <u>N_2 gas</u>. This is of no benefit to living organisms.

7) Some <u>nitrogen-fixing bacteria</u> live in the <u>soil</u>. Others live in <u>nodules</u> on the roots of <u>legume plants</u>. This is why legume plants are so good at putting nitrogen <u>back into the soil</u>. The plants have a <u>mutualistic relationship</u> with the bacteria — the bacteria get <u>food</u> (sugars) from the plant, and the plant gets <u>nitrogen compounds</u> from the bacteria to make into <u>proteins</u>.

A Biosphere Could be Used to Colonise Mars

1) Scientists are able to create <u>artificial biospheres</u> — <u>sealed, self-contained environments</u>. Everything in a biosphere has to be kept in <u>balance</u>, e.g. CO_2, O_2 and food.

2) <u>Mars</u> has a very <u>different environment</u> from Earth, so it's unlikely that organisms from Earth could survive there. However it's been suggested that one way humans could survive on Mars is if they set up an <u>artificial biosphere</u> there — containing similar conditions to those found on Earth.

It's the cyyyycle of liiiiife...

Nitrogen is <u>vital</u> to living things, because it's found in <u>proteins</u>, which are needed for things like enzymes.

Revision Summary for Section Nine

Here goes, folks — another beautiful page of revision questions to keep you at your desk studying hard until your parents have gone out and you can finally nip downstairs to watch TV. Think twice though before you reach for that remote control. These questions are actually pretty good — certainly more entertaining than 'Train Your Husband Like He's a Dog' or 'Celebrities Dance Around'. Question 14 is almost as good as an episode of 'Supernanny'. Question 4 is the corker though — like a reunion episode of 'Friends' but a lot funnier. Give the questions a go. Oh go on.

1) Write down the equation for photosynthesis.

2) Write down five ways that plants can use the glucose produced by photosynthesis.

3)* The graph shows how the rate of plant growth is affected by increasing the level of carbon dioxide.
   Look at the graph and answer the two questions below.

   a) At what level of carbon dioxide is the plant's growth limited by another factor?

   b) Suggest two possible limiting factors on the plant's growth above this level.

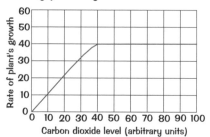

4) What happens to plant enzymes if the temperature is raised over about 45 °C?

5) How does being broad and thin help a leaf to photosynthesise?

6) Why does carbon dioxide tend to move into leaves when they're photosynthesising?

7) Give three ways that the transpiration stream benefits a plant.

8) How is the transpiration rate affected by: a) increased temperature, b) increased air humidity?

9) Stomata close automatically when a plant is short of water. How does this benefit the plant?

10) What is turgor pressure?

11) How are xylem vessels adapted to their function?

12) Name the three main minerals plants need for healthy growth.

13) How can you tell by looking at a plant that it isn't getting enough phosphates?

14) What is active transport? Why is it used in the roots of a plant?

15) Explain why number pyramids are not always pyramid-shaped.

16) What is the source of all the energy in a typical food chain?

17) Why is it unusual to find a food chain with more than five trophic levels?

18) Give three reasons why developing biofuels is a great idea.

19) What is mycoprotein used for?

20) Give the four main ways in which it is possible to maximise food production.

21) Explain how fish farms reduce the following: a) energy loss,   b) disease,   c) threat of predators.

22) Give two advantages and two disadvantages of biological pest control.

23) What is meant by the term hydroponics?

24) Give an example of a detritivore.

25) How does carbon enter the carbon cycle from the air?

26) What important role do nitrogen-fixing bacteria play in the nitrogen cycle?

* Answers on page 140

Solute Exchange — Active Transport

The processes that keep organisms alive won't happen without the right raw materials.
And the raw materials have to get to the right places. It's like making chicken soup.
You need the chicken in your kitchen. It's no good if it's still at the supermarket.

Substances Move by Diffusion, Osmosis and Active Transport

1) Life processes need gases or other dissolved substances before they can happen.

2) For example, for respiration to take place, glucose and oxygen both have to get inside cells.

3) Waste substances also need to move out of the cells so that the organism can get rid of them.

4) These substances move to where they need to be by diffusion, osmosis and active transport.

5) Diffusion is where particles move from an area of high concentration to an area of low concentration — see page 56.

6) Osmosis is similar, but it only refers to water. The water moves across a partially permeable membrane (e.g. a cell membrane) from an area of high water concentration to an area of low water concentration — see page 58.

7) Diffusion and osmosis both involve stuff moving from an area where there's a high concentration of it, to an area where there's a lower concentration of it. Sometimes substances need to move in the other direction — which is where active transport comes in...

We Need Active Transport to Stop Us Starving

Active transport is used in the digestive system when there is a low concentration of nutrients in the gut, but a high concentration of nutrients in the blood.

1) When there's a higher concentration of glucose and amino acids in the gut they diffuse naturally into the blood.

2) BUT — sometimes there's a lower concentration of nutrients in the gut than there is in the blood.

3) This means that the concentration gradient is the wrong way. The food molecules should go the other way if they followed the rules of diffusion.

4) The answer is that a conveniently mysterious process called 'active transport' is responsible.

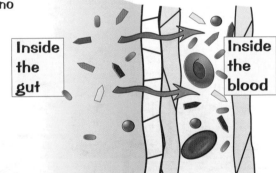

Inside the gut | Inside the blood

5) Active transport allows nutrients to be taken into the blood, despite the fact that the concentration gradient is the wrong way. This is essential to stop us starving. But active transport needs ENERGY from respiration to make it work.

Plants Take in Minerals Using Active Transport

The concentration of minerals is usually higher in the root hair cell of the plant than in the soil around it. So normal diffusion doesn't explain how minerals are taken up into the root hair cell. The same process used in the gut is used here... active transport.

Active transport sucks...

An important difference between active transport and diffusion is that active transport uses energy. Imagine a pen of sheep in a field. If you open the pen, the sheep will happily diffuse from the area of high sheep concentration into the field, which has a low sheep concentration — you won't have to do a thing. To get them back in the pen though, you'll have to put in quite a bit of energy.

The Respiratory System

You need to get <u>oxygen</u> from the air into your bloodstream so that it can get to your cells for <u>respiration</u>. You also need to get rid of <u>carbon dioxide</u> in your blood. This all happens inside the <u>lungs</u>. Breathing is how the air gets <u>in and out</u> of your <u>lungs</u>. Breathing's definitely a useful skill to have. You'll need to be able to do it to get through the exam.

The Lungs Are in the Thorax

1) The <u>thorax</u> is the top part of your 'body'.

2) It's separated from the lower part of the body by the <u>diaphragm</u>.

3) The lungs are like big pink <u>sponges</u>. and are protected by the <u>ribcage</u>.

4) The air that you breathe in goes through the <u>trachea</u>. This splits into two tubes called '<u>bronchi</u>' (each one is 'a bronchus'), one going to each lung.

5) The bronchi split into progressively smaller tubes called <u>bronchioles</u>.

6) The bronchioles finally end at small bags called <u>alveoli</u> where the gas exchange takes place.

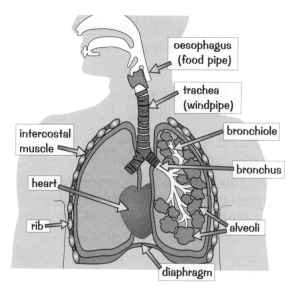

Breathing In...

1) <u>Intercostal muscles</u> and <u>diaphragm contract</u>.
2) Thorax volume <u>increases</u>.
3) This decreases the pressure, drawing air <u>in</u>.

...and Breathing Out

1) <u>Intercostal muscles</u> and <u>diaphragm relax</u>.
2) Thorax volume <u>decreases</u>.
3) Air is forced <u>out</u>.

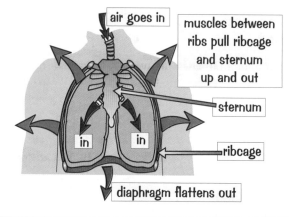

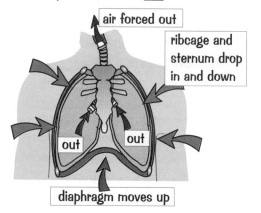

Cilia and Mucus Protect the Lungs

1) The <u>respiratory tract</u> (trachea and bronchi) is lined with <u>mucus</u> and <u>cilia</u> (little hairs) which catch <u>dust</u> and <u>microbes</u> before they reach the lungs.

2) The cilia <u>beat</u>, pushing microbe-filled mucus out of the lungs as phlegm.

3) Sometimes the microbes get past the body's defences and cause infection. The lungs are particularly <u>prone</u> to <u>infections</u> because they're a <u>dead end</u> — microbes can't easily be flushed out.

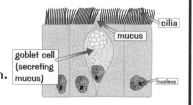

Stop huffing and puffing and just LEARN IT...

So when you breathe in, you don't have to suck the air in. You just make the space in your lungs <u>bigger</u> and the air rushes in to fill it. The small bags called <u>alveoli</u> at the ends of the air passages are the really interesting bit. It's through the alveoli that the <u>oxygen</u> gets into the blood supply to be carted off round the body. Also, the waste <u>carbon dioxide</u> gets out of the blood supply here so it can be breathed out.

Lung Capacity and Disease

If you've ever blown into a white tube at the doctors the chances are you were having your
<u>lung capacity</u> measured — this gives the doctor an idea of how <u>healthy</u> your lungs are.

Lung Capacity — <u>The Total Volume of Air In Your Lungs</u>

1) The <u>total volume of air</u> you can fit in your lungs is your <u>total lung capacity</u> (usually about 6 litres).

2) The volume of air you breathe in (or out) in <u>one normal breath</u> is called your <u>tidal volume</u>.

3) Even if you try to breathe out really hard there's always <u>some air left</u> (just over a litre) in your lungs to make sure that they <u>stay open</u> — this is called the <u>residual volume</u>.

4) Total lung capacity minus residual volume gives you <u>vital capacity</u> — the amount of usable air.

Lung Capacity <u>Can be</u> <u>Measured with a Spirometer</u>

Doctors measure lung capacity using a machine called a <u>spirometer</u> — it can help <u>diagnose</u> and <u>monitor lung diseases</u>.

The patient breathes into the machine (through a tube) for a few minutes, and the volume of air that is breathed in and out is measured and plotted on a graph (called a <u>spirogram</u>) — like this one...

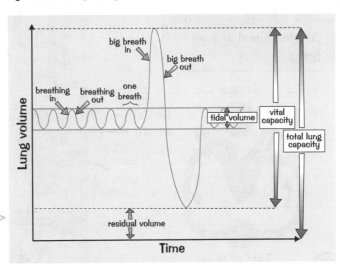

Lung Disease <u>Can be Caused by Lots of Things</u>

1) **Industrial materials** e.g. <u>asbestos</u>. Asbestos can cause cancers, as well as a disease called <u>asbestosis</u> (where lung tissue is scarred, causing breathlessness and even death). Asbestos used to be used as an insulator in roofs, floors, furnaces, etc. Its use is more tightly controlled now.

2) **Genetic causes** e.g. <u>cystic fibrosis</u> is an inherited lung condition. A single defective gene causes the lungs to produce a really thick, sticky <u>mucus</u> that clogs up the lungs — this makes breathing difficult and can lead to life-threatening infections.

3) **Lifestyle causes** e.g. <u>smoking</u> can cause <u>lung cancer</u> (see p.23). This is where <u>cells</u> divide <u>out of control</u>, forming a <u>tumour</u>. The abnormal cells can get into the blood and cause tumours elsewhere.

4) **Asthma** Asthma affects around 1 in 12 adults in the UK. Asthmatics' lungs are <u>overly sensitive</u> to certain things (e.g. pet hair, pollen, dust, smoke...). When they encounter these things the <u>muscles</u> around the airways <u>constrict</u>, narrowing the airways and making it hard to breathe (an <u>asthma attack</u>). Symptoms of an attack are <u>shortness of breath</u>, <u>coughing</u>, <u>wheezing</u> and a <u>tight chest</u>. When symptoms appear a muscle relaxant drug is inhaled (from an <u>inhaler</u>) to open up the airways. Some people also take drugs to stop attacks happening in the first place (but there's <u>no actual cure</u>.)

Spirograms... aren't they those fancy drawing machines...

If the values on a spirogram are <u>low</u> the person might have a <u>lung disease</u>. If the <u>tidal volume</u> increases (i.e. if they're <u>breathing deeper</u>), then they're probably <u>exercising</u>. A spirogram can also be used to calculate <u>breathing rate</u> (by counting the number of breaths in a minute). Simple really.

The Circulation System

There wouldn't be much point having lungs if you couldn't deliver the oxygen to where it needed to be. This is where the circulation system comes in, its main function is to get <u>food and oxygen</u> to every cell in the body.

The DOUBLE Circulation System, Actually

① The <u>heart</u> is actually <u>two pumps</u>. The <u>right side</u> pumps deoxygenated blood to the <u>lungs</u> to <u>collect oxygen</u> and <u>remove carbon dioxide</u>.
Then the <u>left side</u> pumps this oxygenated blood <u>around the body</u>.

② <u>Arteries</u> carry blood <u>away from the heart</u> at <u>high pressure</u>.

③ Normally, arteries carry <u>oxygenated blood</u> and veins carry <u>deoxygenated blood</u>.

The <u>pulmonary artery</u> and <u>pulmonary vein</u> are the <u>big exceptions</u> to this rule (see diagram).

④ The arteries eventually split off into thousands of tiny <u>capillaries</u> which take blood to <u>every cell</u> in the body.

⑤ The <u>veins</u> then collect the "used" blood and carry it <u>back to the heart</u> at <u>low pressure</u> to be pumped round again.

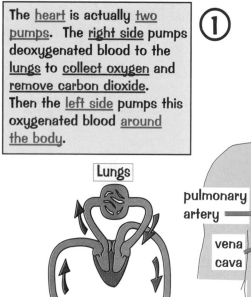

Lungs

Rest of Body

brain
aorta
lungs
pulmonary artery
pulmonary vein
vena cava
heart
liver
gut
kidneys
from lower limbs
to lower limbs

The Cardiac Cycle is How the Heart Contracts

The sequence of events in <u>one complete heartbeat</u> is called the <u>cardiac cycle</u>:

① <u>Blood flows into</u> the two <u>atria</u>.

② The <u>atria contract</u>, pushing the blood into the <u>ventricles</u>.

③ The <u>ventricles contract</u>, forcing the blood into the <u>aorta</u> and the <u>pulmonary artery</u>.

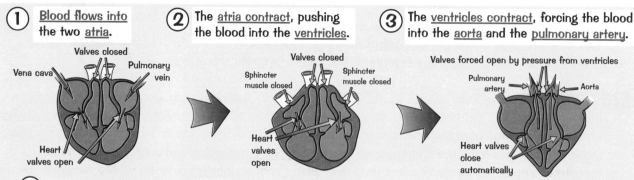

④ The blood then flows along the arteries, the atria fill again and the whole <u>cycle</u> starts over.

Blood vessels — a vampire's favourite type of ship...

The diagram above only shows the <u>basic layout</u>. There's actually <u>zillions</u> of blood vessels. If you laid all your arteries, capillaries and veins end to end, they'd go around the world about three times. These vessels vary from hose-pipe width <u>arteries</u> to <u>capillaries</u> that are a tenth of the thickness of a human hair.

The Heart and Heart Disease

The heart is pretty simple, it's a pump that pushes blood around the body, but if something goes wrong the consequences can be dire.

The Heart Has a Pacemaker

1) The heart is told <u>how fast to beat</u> by a group of cells called the <u>pacemaker cells</u>.

2) These cells produce a small <u>electric current</u> which spreads to the surrounding muscle cells, causing them to <u>contract</u>.

3) There are <u>two</u> clusters of these cells in the heart:
   The <u>sino-atrial node</u> (SAN) stimulates the <u>atria</u> to contract.
   The <u>atrio-ventricular node</u> (AVN) stimulates the <u>ventricles</u> to contract.

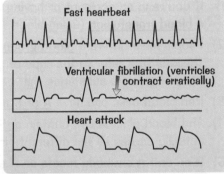

4) In one complete heartbeat the SAN produces an electric current <u>first</u>, which spreads to the atria (making them contract). The current stimulates the AVN to produce an electric current (causing the ventricles to contract). This process ensures that the <u>atria always contract before the ventricles</u>.

5) An <u>artificial pacemaker</u> can be used to control heartbeat if the pacemaker cells don't work properly. It's a little device that's implanted under the skin and has a wire going to the heart. It produces an electric current.

ECGs and Echocardiograms Measure the Heart

Doctors can measure how well the heart is working (<u>heart function</u>) in two main ways.

1) <u>Electrocardiogram</u> (ECG) — showing the <u>electrical activity</u> of the heart. They can show:
   • <u>heart attacks</u> — e.g. if you're having a heart attack, or are about to have one,
   • <u>irregular heartbeats</u> and <u>general health</u> of the heart.

This is what a healthy person's ECG looks like...   ...and here are some unhealthy ones.

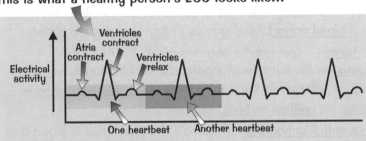

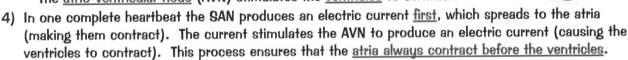

2) <u>Echocardiogram</u> — an <u>ultrasound scan</u> of the heart,
   which can show: • enlarged heart — this could indicate heart failure,
                   • decreased pumping ability — this could indicate a disease called cardiomyopathy,
                   • valve function — torn, infected or scarred heart valves can cause problems.

Lifestyle Affects the Health of the Heart

<u>Heart disease</u> is when the arteries that supply blood to the muscle of the heart get <u>blocked</u> by fatty deposits. This often results in a <u>heart attack</u>. There are several common lifestyle 'risk factors' for heart disease...

1) Unhealthy diet
2) Drinking alcohol
3) Smoking
4) Stress
5) Drugs

A stitch — the best running pacemaker in the world...

In the exam you might be asked to <u>interpret</u> an ECG — they look scary but they're not too difficult... If a <u>peak</u> is <u>missing</u>, then that part of the heart <u>isn't contracting</u>. If the peaks are <u>close together</u>, the heart's beating <u>faster</u>. But if everything is going <u>haywire</u>, then it could be a <u>heart attack</u> or <u>fibrillation</u>.

Blood

If you get a cut, you don't want all your blood to drain away — this is why <u>clotting</u> is so handy. Sometimes injuries are so bad you lose a lot of blood and you need to replace it — that's where <u>transfusions</u> come in. [The structures and functions of blood cells are covered on page 63.]

Blood Sometimes Doesn't Clot Properly

1) When you're injured, your blood <u>clots</u> to <u>prevent too much bleeding</u>. <u>Platelets</u> clump together to 'plug' the damaged area. In a clot, platelets are held together by a mesh of a protein called <u>fibrin</u> (though this process also needs other proteins called <u>clotting factors</u> to work properly).

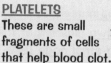

PLATELETS
These are small fragments of cells that help blood clot.

2) Some substances in food (and drink) affect the way the blood clots:
   • <u>Vitamin K</u> — this is needed for blood to clot properly. <u>Green vegetables</u> contain lots of vitamin K.
   • <u>Alcohol</u> — a moderate intake of alcohol slows blood clotting.
   • <u>Cranberries</u> — it's been suggested that they may slow blood clotting (but more research is needed).

3) Too little clotting could mean you bleed to death (well, you're more likely get loads of bruises). Too much clotting can cause <u>strokes</u> and <u>deep vein thrombosis</u> (DVT).

4) People who are at risk of stroke and DVT can take <u>drugs</u> to reduce their risk. <u>Warfarin</u>, <u>heparin</u> and <u>aspirin</u> all help <u>prevent</u> the blood from clotting.

5) <u>Haemophilia</u> is a <u>genetic condition</u> where the blood <u>doesn't clot easily</u> because a <u>clotting factor</u> can't be made by the body — this missing clotting factor can be injected.

Blood Type is Important in Transfusions

1) If you're in an accident or having surgery, you may lose a lot of blood — this needs to be replaced by a <u>blood transfusion</u> (using blood from a <u>blood donor</u>). But you can't just use any old blood...

2) People have different <u>blood groups</u> or <u>types</u> — you can be any one of: A, B, O or AB. These letters refer to the type of <u>antigens</u> on the surface of a person's red blood cells. (An antigen is a substance that can trigger a response from a person's <u>immune system</u>.)

3) Red blood cells can have <u>A or B antigens</u> (or <u>neither</u>, or <u>both</u>) on their surface.

4) And blood plasma can contain <u>anti-A or anti-B antibodies</u>. (Plasma's the pale liquid in blood that actually carries all the different bits — e.g. the blood cells, antibodies, hormones, etc.)

5) If an anti-A antibody meets an A antigen, the blood clots up and it all goes <u>hideously wrong</u>. Same thing when an anti-B antibody meets a B antigen. (This is <u>agglutination</u> — a fancy name for 'clumping together'. The <u>antibodies</u> are acting as <u>agglutinins</u> — or 'things that make stuff clump together'.)

6) This table should make everything lovely and clear...

Blood Group	Antigens	Antibodies	Can give blood to	Can get blood from
A	A	anti-B	A and AB	A and O
B	B	anti-A	B and AB	B and O
AB	A, B	none	only AB	anyone
O	none	anti-A, anti-B	anyone	only O

For example, '<u>O blood</u>' can be given to <u>anyone</u> — there are <u>no antigens</u> on the blood cells, so any <u>anti-A</u> or <u>anti-B antibodies</u> have nothing to 'attack'.

I think I need an information transfusion... from this book to my brain...

You might get asked a question on <u>who</u> can donate blood to <u>who</u> (or vice versa) in the exam. Just look at what blood type the donor is and think about what <u>antigens</u> and <u>antibodies</u> they have in their blood. It's <u>hard</u>, and you need to think <u>carefully</u> about it (I do anyway), but it does make sense.

Waste Disposal — The Kidneys

The kidneys are really important organs. They get rid of toxic waste like urea as well as adjusting the amount of dissolved ions and water in the blood. The kidneys were introduced on page 66, but here's the rest of the stuff you need to know.

Nephrons Are the Filtration Units in the Kidneys

1) Ultrafiltration:

1) A high pressure is built up which squeezes water, urea, ions and sugar out of the blood and into the Bowman's capsule.

2) The membranes between the blood vessels and the Bowman's capsule act like filters, so big molecules like proteins and blood cells are not squeezed out. They stay in the blood.

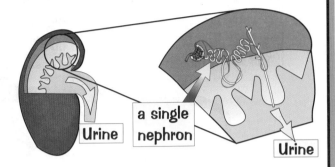

Enlarged View of a Single Nephron

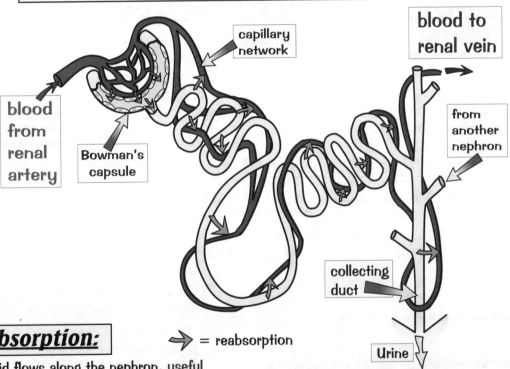

2) Reabsorption:

$\Rightarrow$ = reabsorption

As the liquid flows along the nephron, useful substances are reabsorbed back into the blood:

1) All the sugar is reabsorbed. This involves the process of active transport against the concentration gradient.

2) Sufficient ions are reabsorbed. Excess ions are not. Active transport is needed.

3) Sufficient water is reabsorbed.

3) Release of wastes:

The remaining substances (including urea) continue out of the nephron, into the ureter and down to the bladder as urine.

Don't try to kid-me that you know it all — learn it properly...

The kidneys are pretty complicated organs as you can see. Luckily you don't have to learn all the ins and outs of the diagram — but you do have to make sure you know exactly what happens in each of the three stages. Learn what's filtered, what's reabsorbed and what's released as urine.

Waste Disposal — The Kidneys

Water Content is Controlled by the Kidneys

1) The amount of water reabsorbed in the kidney nephrons is <u>controlled</u> by a hormone called <u>anti-diuretic hormone</u> (ADH).

2) The brain <u>monitors the water content of the blood</u> and instructs the <u>pituitary gland</u> to release <u>ADH</u> into the blood according to how much is needed.

3) The whole process of water content regulation is controlled by a mechanism called <u>negative feedback</u> (see page 11). This means that if the water content gets <u>too high</u> or <u>too low</u> a mechanism will be triggered that brings it back to <u>normal</u>.

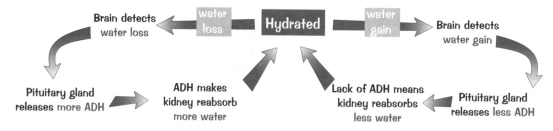

The Kidneys Remove Waste Substances from the Blood

1) If the kidneys don't work properly, <u>waste substances build up</u> in the blood and you lose your ability to <u>control</u> the <u>levels of ions and water</u> in your body. Eventually, this results in <u>death</u>.

2) People with kidney failure can be kept alive by having <u>dialysis treatment</u> — where <u>machines</u> do the job of the kidneys. Or they can have a <u>kidney transplant</u>.

> The kidneys are incredibly important — if they don't work as they should, you can get problems in the <u>heart</u>, <u>bones</u>, <u>nervous system</u>, <u>stomach</u>, <u>mouth</u>, etc.

Dialysis Machines Filter the Blood

1) Dialysis has to be done <u>regularly</u> to keep the concentrations of <u>dissolved substances</u> in the blood at <u>normal levels</u>, and to remove waste substances.

2) In a <u>dialysis machine</u> the person's blood flows alongside a <u>selectively permeable barrier</u>, surrounded by dialysis fluid. It's permeable to things like <u>ions</u> and <u>waste substances</u>, but not <u>big molecules</u> like proteins (just like the membranes in the kidney).

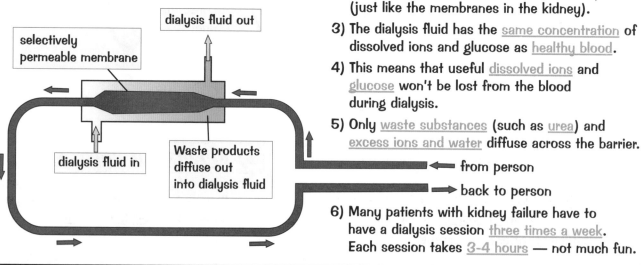

3) The dialysis fluid has the <u>same concentration</u> of dissolved ions and glucose as <u>healthy blood</u>.

4) This means that useful <u>dissolved ions</u> and <u>glucose</u> won't be lost from the blood during dialysis.

5) Only <u>waste substances</u> (such as <u>urea</u>) and <u>excess ions and water</u> diffuse across the barrier.

6) Many patients with kidney failure have to have a dialysis session <u>three times a week</u>. Each session takes <u>3-4 hours</u> — not much fun.

Simon says touch urea... actually don't...

Kidney failure patients often have <u>high blood pressure</u> because diseased kidneys can't control the <u>water content</u> of the <u>blood</u>. This excess water is removed during dialysis (up to 5 litres of fluid can be removed in one session). Dialysis is normally done three times a week for about 4 hours at a time — not fun.

Organ Replacements and Donation

If an organ's severely damaged, it can be replaced by an artificial part or a donated natural organ. Wow.

Organs Can be Replaced by Living or Dead Donors

1) Living donors can donate whole (or parts of) certain organs. For example, you can live with just one of your two kidneys and donate the other, or you can donate a piece of your liver. To be a living donor you must be fit and healthy, over 18, and usually a close family member (for a good tissue match).

2) Organs from people who have recently died, or who are brain dead, can also be transplanted.

3) Any donor organ must by relatively young, the right size and a good tissue match.

4) But there's a big shortage of donors...

> The UK has a shortage of organs available for donation.
>
> - You can join the NHS Organ Donor Register to show you're willing to donate organs after you die. However, doctors still need your family's consent before they can use the organs for a transplant.
> - Some people say it should be made easier for doctors to use the organs of people who have died. One suggestion is to have an 'opt-out' system instead — this means anyone's organs can be used unless the person has registered to say they don't want them to be donated.

5) Success rates of transplants depend on a lot of things — e.g. the type of organ (e.g. the heart is riskier than a kidney), the age of the patient, the skill of the surgeon, etc.

6) But transplants involve major surgery — and even if all goes well, there can be problems with rejection or taking immunosuppressive drugs (see p.65).

There are Issues Surrounding Organ Donation

Like with lots of medical advances, there are ethical issues...

1) Some people think for religious reasons that a person's body should be buried intact (so giving organs is wrong). Others think life or death is up to God (so receiving organs is wrong).

2) Others worry that doctors might not save them if they're critically ill and their organs are needed for transplant. There are safeguards in place that should prevent this though.

3) There are also worries that people may get pressured into being a 'living donor' (e.g. donating a kidney to a close relative). But doctors try to ensure that it's always the donor's personal choice.

Mechanical Replacements can Sometimes be Used

1) Mechanical (artificial) replacements made of metal and plastic can also be used. These don't have the same problems with rejection, but they have a whole new set of problems instead. For example, artificial heart valves need more major surgery and don't work as well as healthy natural ones.

2) Sometimes, temporary mechanical replacements are needed to keep someone alive. This could be for anything from a few hours (e.g. during an operation), to several months or even years (e.g. if they're waiting for a suitable organ donor). For example...

> - A heart-lung machine keeps a patient's blood oxygenated and pumping during heart or lung surgery.
> - A kidney dialysis machine can filter a patient's blood (e.g. while they wait for a kidney transplant). See p.100 for more info.

I think I need a brain transplant to learn all this lot...

Did you know... that if you transplant a piece of a liver it can actually grow back to normal size within a few weeks. Impressive. Changing the subject slightly... one donor can donate several organs — e.g. their heart, kidneys, liver, lungs, pancreas... And on top of that, other tissues (e.g. skin, bone, tendons, corneas...) can also be donated. It's absolutely amazing really, when you think about it.

Bones and Cartilage

Bones and joints are pretty darned important — without them you wouldn't be able to move around at all. All you'd do is wobble around on the floor like a big squidgy thing.

If You Didn't Have a Skeleton, You'd be Jelly-like

1) The job of a skeleton is to support the body and allow it to move — as well as protect vital organs.

2) Fish, amphibians, reptiles, birds and mammals are all vertebrates — they all have a backbone and an internal skeleton. Other animals (e.g. insects) have their skeleton on the outside.

3) An internal skeleton has certain advantages
- It can easily grow with the body.
- It's easy to attach muscles to it.
- It's more flexible than an external skeleton.

Bones are Living Tissues...

Bones are a lot cleverer than they might look...

1) Bones are made up of living cells — so they grow, and can repair themselves if they get damaged.

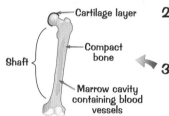
Cartilage layer
Compact bone
Shaft
Marrow cavity containing blood vessels

2) Long bones (e.g. the big one in your thigh) are actually hollow — this makes them lighter than solid bones of the same size (and stronger than solid bones of the same mass). This makes movement far more efficient.

3) The hole in the middle of some long bones is filled with bone marrow. Bone marrow is a spongy substance that makes new blood cells — meaning your bones are actually a kind of blood factory.

...That Start Off Life as Cartilage

1) Bones start off as cartilage in the womb. (Cartilage is living tissue that looks and feels a bit rubbery.)

2) As you grow, cartilage is replaced by bone. Blood vessels deposit calcium and phosphorus in the cartilage — which eventually turns it into bone. This process is called ossification.

3) You can tell if someone is still growing by looking at how much cartilage is present — if there's a lot, they're still growing.

4) Even when you're fully grown, the ends of bones remain covered with cartilage (to stop bones rubbing together at joints — see next page).

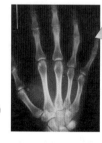
Bones show up on an X-ray, but cartilage doesn't.

X-rays can also show where fractures are.

Bones and Cartilage Can Get Damaged

1) Cartilage and bone are both made up of living tissue, and so can get infected. (The top of the ear is made of cartilage — if you get this pierced, you have to make sure no infection gets in. Not nice.)

2) Even though bones are really strong, they can be fractured (broken) by a sharp knock. Elderly people are more prone to breaking bones as they often suffer from osteoporosis — a condition where calcium is lost from the bones. (Osteoporosis makes the bones softer, more brittle and more likely to break — it can be treated with calcium supplements.)

3) A broken bone can easily injure nearby tissue — so you shouldn't move anyone who might have a fracture. That's especially true for someone with a suspected spinal fracture (broken back) — moving them could damage their spinal cord (basically an extension of the brain running down the middle of the backbone). Damage to the spinal cord can lead to paralysis.

No bones about it... it's a humerus page...

Bones are all too easily thought of as just organic scaffolding. But they're pretty amazing really, and painful if you break one. Talking of broken bones... there are different kinds of break. You've got simple fractures, compound fractures (where the bone pokes through the skin), green-stick fractures... and so on. But bones usually mend pretty easily — if you hold them still, a break will knit itself together.

Joints and Muscles

Like it says in the song, the knee bone's connected to the thigh bone. And it's done with a joint. Read on.

Joints Allow the Bones to Move

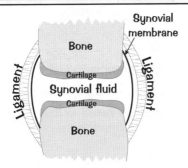

1) The bones at a joint are held together by <u>ligaments</u>. Ligaments have <u>tensile strength</u> (i.e. you can pull them and they don't snap easily) but are pretty <u>elastic</u> (stretchy).

2) The ends of bones are covered with <u>cartilage</u> to stop the bones <u>rubbing</u> together. And because cartilage can be slightly compressed, it can act as a <u>shock absorber</u>.

3) <u>Membranes</u> at joints release oily <u>synovial fluid</u> to <u>lubricate</u> the joints, allowing them to move more easily.

4) Different kinds of joints move in different ways. For example...

BALL AND SOCKET ...like the <u>hip</u> or <u>shoulder</u>. The joint can move in <u>all</u> <u>directions</u>, and can also <u>rotate</u>.

HINGE ...like the <u>knee</u> or <u>elbow</u>. The joint can go <u>backwards and</u> <u>forwards</u>, but not side-to-side.

Muscles Pull on Bones to Move Them

1) Bones are attached to muscles by <u>tendons</u>.

2) Muscles move bones at a joint by <u>contracting</u> (becoming <u>shorter</u>). They can only <u>pull</u> on bones to move a joint — they <u>can't</u> push.

3) This is why muscles usually come in <u>pairs</u> (called <u>antagonistic pairs</u>). When one muscle in the pair contracts, the joint moves in one direction. When the other contracts, it moves in the <u>opposite</u> direction.

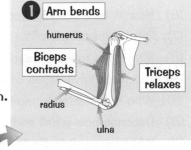

❶ Arm bends
humerus
Biceps contracts
Triceps relaxes
radius
ulna

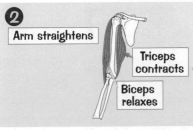

❷ Arm straightens
Triceps contracts
Biceps relaxes

4) The <u>biceps</u> and <u>triceps</u> are an antagonistic pair of muscles. When the <u>biceps</u> contracts it pulls the lower arm <u>upwards</u>. And when the <u>triceps</u> contracts the lower arm is pulled back <u>down</u>.

5) Together, they make the arm work as a <u>lever</u>, where the elbow is the pivot.

Joints Can be Replaced

If your <u>hip</u> or <u>knee joints</u> get damaged or diseased, they can be <u>replaced</u> with <u>artificial joints</u>. Assuming all goes well, you'll be in <u>less pain</u> and discomfort, and <u>able to walk</u> better. But there are <u>disadvantages</u>...

1) The surrounding tissue may become <u>inflamed</u> and <u>painful</u> — this is caused by the body's reaction to the material the joint is made of.

2) <u>Hip dislocation</u> (ball comes out of its socket) is more common with artificial joints, as are <u>blood clots</u>.

3) There's a <u>risk of infection</u>, as with any surgery.

4) The <u>length of the legs</u> may be slightly different, causing difficulty walking.

5) Artificial joints <u>don't last forever</u> — they usually have to be replaced after 12-15 years.

What's a skeleton's favourite instrument?... a trom-bone...

Different joints have different <u>ranges of movement</u>. And if you do something that makes the bone move further than its range of movement (like fall on it), then you could <u>dislocate</u> it. Painful.

104

Revision Summary for Section Ten

It's no good just reading the section through and hoping you've got it all — it'll only stick if you've learned it <u>properly</u>. These questions are designed to really test whether you know all your stuff — ignore them at your peril. OK, rant over — I'll leave it to you...

1) Give the two main differences between active transport and diffusion.
2) Why can't most mineral ions get into roots by diffusion?
3) Describe what happens to the intercostal muscles and diaphragm when you breathe in and out.
4) What machine would a doctor use to measure lung capacity? Why would you want to measure it?
5) List the four main causes of lung disease. Which of these is the most likely to result in lung cancer?
6) Describe the circulatory system of humans.
7) Describe the four stages that make up one complete heart beat.
8) Name the two clusters of pacemaker cells in the heart. What do they do?
9) What does ECG stand for? Describe what a healthy person's ECG should look like.
10) List five lifestyle 'dangers' which may cause an increased likelihood of cardiovascular disease.
11) What exactly is happening when blood clots? How might your diet affect the way your blood clots?
12) Explain what would happen if a person with type A blood was given a transfusion of type B blood.
13) Explain how a kidney works.
14) Describe three things that affect the amount and concentration of urine produced.
15) Which hormone is responsible for controlling the amount of water reabsorbed in the kidneys?
16) How does a dialysis machine work? Which substances does it remove from the blood?
17) Give one example of an ethical concern surrounding organ donation.
18) Name two types of mechanical organ replacements. What do they do?
19) List three advantages of an internal skeleton over an external skeleton.
20) Why is it unwise to move someone with a broken bone?
21) Describe how a ball and socket joint works.
22) What happens to the lower arm when the triceps contracts?

Section Ten — Organs and Systems 2

Bacteria

Bacteria are a type of microorganism. They're tiny — typical bacteria are just a few microns (thousandths of a millimetre) wide. But despite their small size, they can have a mighty effect on humans...

Bacterial Cells are Usually Smaller and Simpler than Animal Cells

1) This table shows how bacterial cells compare to plant and animal cells.

2) Bacterial cells don't have a proper nucleus like plant and animal cells do. They have bacterial DNA to control the cell's activities and replication, but the DNA just floats about in the cytoplasm.

3) They also don't have any mitochondria, chloroplasts or a vacuole.

4) They have a cell wall to keep their shape and stop them bursting. This isn't the same kind of cell wall found in a plant though.

Feature	Animal Cell	Plant Cell	Bacterial Cell
Nucleus	✓	✓	✗
Cell membrane	✓	✓	✓
Mitochondria	✓	✓	✗
Cell Wall	✗	✓	✓
Chloroplasts	✗	✓	✗
Vacuole	✗	✓	✗
Extras	None	None	Flagellum

5) They sometimes have a flagellum (tail) to help them move.

6) They come in 4 shapes: rods, curved rods, spheres and spirals.

7) Bacteria can consume a huge range of organic nutrients from their surroundings. This provides them with energy. Some types of bacteria can even produce their own nutrients.

8) This means they can survive pretty much anywhere — in soil, water, air, in your house, in the human body and in food.

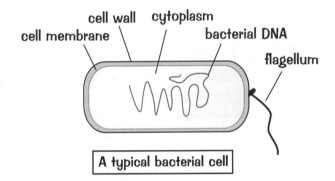

cell wall   cytoplasm
cell membrane   bacterial DNA
                                    flagellum

A typical bacterial cell

Bacteria Reproduce by Asexual Reproduction

1) Bacteria reproduce by asexual reproduction — they're clones of each other. They reproduce by a process called binary fission (a posh way of saying 'they split in two').

2) Bacteria reproduce very quickly. If disease-causing bacteria enter your body, they can reproduce and cause disease before your body has a chance to respond.

3) Bacteria reproduce more quickly in certain conditions. Generally, if it's warm and they have a good source of nutrients then they will grow better. This is why it's important to store food carefully. If you leave some meat on a warm kitchen top, bacteria on the meat will reproduce very quickly and cause it to spoil (go off). But if you put the same meat in the fridge, then the cold temperature will slow down the bacteria's reproduction and it won't spoil as quickly.

Bacteria aren't all bad though...

Not all types of bacteria make you ill or spoil your food. Bacteria are also used to make foods like yoghurt and soy sauce (see p.107), and food additives like Vitamin C and MSG (see p.108). Bacteria that can make Belgian chocolates, caviar and profiteroles haven't been discovered yet...

Harmful Microorganisms

There are different kinds of microorganism, e.g. bacteria, viruses, fungi and protozoa. Some are useful, while others are pretty harmful if you get infected. This page focuses on the nasty ones...

There Are Four Stages in an Infectious Disease

1 Firstly the microorganism has to get into the body to cause an infection. There are four main ways they can do it:

- Through the nose, e.g. airborne microorganisms like the influenza virus are breathed in.
- Through the mouth, e.g. contaminated food and water causes food poisoning and cholera.
- Through the skin, e.g. cuts, insect bites and infected needles can introduce a pathogen into the body directly or they can breach the skin, allowing microorganisms on the skin to get in.
- Through sexual contact, e.g. the HIV virus that causes AIDS can get into the body this way.

2 Once the microorganism is in the body it reproduces rapidly, producing many more microorganisms.

3 The microorganisms then produce toxins (poisonous substances) which damage cells and tissues.

4 The toxins cause symptoms of infection, e.g. pain, diarrhoea and stomach cramps. Your immune system's reaction to the infection can also cause symptoms, e.g. fever. The time between exposure to the microorganism and the development of symptoms is called the incubation period.

Poor Sanitation is Linked to a High Incidence of Disease

1) The incidence of a disease is the number of new cases that occurs in a population in a certain time.
2) Good sanitation and public health measures are linked to a low incidence of disease. A clean water supply, good sewage works, public health education and clean hospitals prevent the spread of disease.
3) Poor sanitation is linked with a high incidence of disease. E.g. a high incidence of food poisoning, dysentery and cholera might be caused by a lack of clean water or a knackered sewage system. A high incidence of septicaemia (a bacterial blood infection) might be caused by poor hygiene in hospital operating theatres or a lack of education about cleaning cuts properly.
4) Developing countries are less likely to be able to afford good sanitation and public health measures.

Diseases Often Spread Rapidly After Natural Disasters

Natural disasters like earthquakes and hurricanes can damage the infrastructure of an area, and completely disrupt health services. In these conditions disease can spread rapidly among the population.

1) Some natural disasters damage sewage systems and water supplies. This can result in contaminated drinking water containing the microorganisms that cause diseases like cholera and dysentery.
2) Transport systems can be damaged — making it difficult for health services to reach people in need.
3) Electricity supplies are also often damaged by natural disasters. This means that food goes off quickly because refrigerators can't work — this can lead to an increase in food poisoning.

Incidence of revision is increasing due to exams...

The body has loads of different defence mechanisms to defend itself against microorganisms. There's things like the skin, mucus and cilia in the respiratory system and white blood cells, see page 24 for more.

Microorganisms and Food

Biotechnology ('bio' meaning life, and 'technology' meaning, well... technology) is nothing new.
For years the food industry has been using microorganisms to produce cheese, yoghurt, chocolate,
soy sauce, alcohol, etc. And more recently we've started using mycoproteins (see page 86).

Bacteria Ferment Milk to Produce Yoghurt...

Fermentation is when microorganisms break sugars down to release energy — usually by
anaerobic respiration. Yoghurt is basically fermented milk. Here's how it's made...

1) The equipment is sterilised to kill off any unwanted microorganisms.

2) The milk is pasteurised (heated up to 72 °C for 15 seconds)
   — again to kill any harmful microorganisms. Then the milk's cooled.

3) A starter culture of bacteria is added, and the mixture incubated
   (heated to about 40 °C) in a vessel called a fermenter (see p.86).

4) The bacteria ferment the lactose sugar in the milk to form lactic acid. Lactic acid causes the milk to
   clot, and solidify into yoghurt (a sample is taken at this stage to make sure it's the right consistency).

5) Finally, flavours (e.g. fruit) and colours are sometimes added and the yoghurt is packaged.

...and Soy Sauce is Made by Fermentation As Well

There are different kinds of soy sauce, and they're generally quite complicated to make.
This is the process behind one particular kind of soy sauce — it involves three different kinds of microbes.

1) Cooked soy beans and roasted wheat are mixed together.

2) The mixture is fermented by the Aspergillus fungus.

3) The mixture is fermented again by yeasts.

4) The mixture is fermented yet again by the Lactobacillus bacterium.

5) The liquid is filtered to remove any gungy bits.

6) Then it's pasteurised to kill off the microorganisms, and finally put into sterile bottles.

Functional Foods are Marketed as Having Health Benefits

A functional food is one that has some kind of health benefit beyond basic nutrition. For example, it might
prevent some kind of disease, or it might (as the marketing folk may put it) 'promote your well-being'.

Plant Stanol Esters

1) Plant stanol esters are chemicals that can lower blood cholesterol and reduce the risk of heart disease.

2) Some food manufacturers add them to spreads and some dairy products. People who are worried
   about their blood cholesterol levels may choose these spreads over normal spreads.

3) Stanols occur naturally in plants, but in very small quantities. Stanols are produced commercially by
   using bacteria to convert sterols (types of fat found in plants like the soya bean) into stanols.

Prebiotics

Some people take substances called prebiotics to promote the growth of 'good' bacteria in the gut.

1) Prebiotics are carbohydrates such as oligosaccharides. They're a food supply for 'good' bacteria.
   'Bad' bacteria and humans can't digest the prebiotic — they don't have the right enzymes.

2) Prebiotics occur naturally in foods like leeks, onions and oats, but you can't get enough of them
   in a normal diet to cause a significant effect. This is why some people take supplements.

While you're in there, make yourself useful...

We all have bacteria in our guts. The 'bad' bacteria can cause disease, but the 'good' bacteria help
digestion. Don't get prebiotics and probiotics mixed up — probiotics are actually bacteria that you eat
to 'top up' the levels in your gut. Interesting fact... there can be up to 2 kg of bacteria in your gut.

Microorganisms and Food

Unfortunately for you yoghurt and soy sauce are just the tip of the iceberg...

Lots of Microbial Products are Used in Food

In the first two examples, the microbial product is used to produce a food.

ENZYMES

1) Enzymes such as invertase are used in the manufacture of sweets and other foods.

2) Invertase converts the sucrose (a sugar) into glucose and fructose (different types of sugar) which taste sweeter. This means that less sugar is needed for the same sweetness — meaning manufacturers can save money and produce lower-calorie sweet foods.

3) Invertase is naturally produced by a yeast called Saccharomyces cerevisiae.

CHYMOSIN

1) Cheese is made using a substance called rennet.

2) Rennet is traditionally obtained from the lining of a calf's stomach and contains an enzyme called chymosin, which clots the milk.

3) But vegetarians probably don't want to eat cheese made with rennet from animals...
...so vegetarian cheese is made using chymosin from genetically modified microorganisms (p43).

4) Basically, the genes responsible for chymosin were isolated from calf stomach cells and put into yeast cells. These were then grown on an industrial scale to produce chymosin.

In these four examples, the microbial product is eaten.

VITAMIN C

1) Vitamin C is used as a dietary supplement (in vitamin pills). It's also added to drinks (and other things, e.g. bread) to stop them going off.

2) A bacterium called Acetobacter naturally produces a chemical easily converted to vitamin C. It's used to produce vitamin C commercially, as it's cheaper and easier than extracting vitamin C from fruit.

CITRIC ACID

1) Citric acid is a flavouring and preservative added to fruit-flavoured fizzy drinks.

2) It's found naturally in citrus fruits — but since fizzy drinks rarely use fresh fruit, citric acid has to be added separately.

3) A fungus called Aspergillus niger is used to commercially produce citric acid.

MONOSODIUM GLUTAMATE (MSG)

1) Monosodium glutamate is a flavour enhancer added to loads of foods (including many Asian foods).

2) It's made from glutamic acid, which is produced by the bacterium Corynebacterium glutamicum.

3) The bacteria secrete glutamic acid (an amino acid) into the medium they're grown in. The glutamic acid is then used to make monosodium glutamate (a sodium salt).

CARRAGEENAN (PRODUCED FROM THE SEAWEED CARRAGEEN)

1) You also need to know about carrageenan — it's a gelling agent that's extracted from the seaweed carrageen.

2) It's also used as an emulsifier in ice cream, jellies, soups and confectionery.

Seaweeds are algae. Some algae are microbes, but others (e.g. seaweed) aren't. So this example is a bit different to the others — no microorganisms are involved.

It all seemed sensible till we got to the seaweed ice cream...

Microbes are used for other things too — these are just examples. Most of the examples above are natural. The chymosin is different — it's made by genetically modified (GM) yeast.

Yeast

This page is all about yeast — a pretty useful microorganism. It helps us make wine and beer.

Yeast is a Microorganism

1) Yeast is a type of fungus.
2) It reproduces asexually by a process called budding. A bulge forms on part of the cell and it eventually becomes a daughter cell, identical to the parent.
3) Yeast can be easily stored in a dry condition — e.g. baker's yeast is dry granules.

vacuole | daughter yeast cell budding off
cell membrane
cytoplasm
cell wall | nuclei containing DNA

Yeast Can Respire Anaerobically or Aerobically

When yeast respires anaerobically (without oxygen) it produces ethanol, carbon dioxide and energy.
This process is called fermentation. Here is the equation for fermentation:

$$\text{glucose} \rightarrow \text{ethanol} + \text{carbon dioxide (+ energy)}$$
$$C_6H_{12}O_6 \rightarrow 2C_2H_5OH + 2CO_2 \text{ (+ energy)}$$

Ethanol is a type of alcohol.

Yeast can also respire aerobically (with oxygen). This releases more energy than anaerobic respiration.
Aerobic respiration is the same for yeast as it is for plants and animals:

$$\text{glucose} + \text{oxygen} \rightarrow \text{carbon dioxide} + \text{water (+ energy)}$$

Whether the yeast respire aerobically or anaerobically depends on whether there is oxygen present.
If oxygen is present it respires aerobically. If oxygen runs out it switches to anaerobic respiration.

Yeast's Growth Rate Varies Depending on the Conditions

The faster yeast respires, the faster it's able to reproduce. The speed (rate) that yeast respires and reproduces varies depending on factors like: the temperature, amount of glucose, level of toxins and pH.

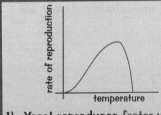

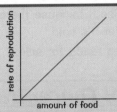

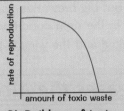

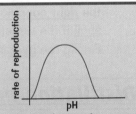

1) Yeast reproduces faster when it's warmer (growth rate doubles for every 10 °C rise in temperature). But if it's too hot the yeast dies.

2) The more food (glucose) there is, the faster the yeast reproduces.

3) Build-up of toxic waste products, e.g. ethanol, slows down reproduction.

4) The pH has to be just right. Too high or low a pH slows down reproduction.

One way of measuring how fast the yeast is reproducing is to measure how much glucose (sugar) it uses up. The faster the yeast reproduces, the more glucose will be used up.

Wastewater Can be Cleaned Up with Yeast

1) Food-processing factories need to get rid of sugary water. They can't just release it into waterways because it would cause pollution. Bacteria in the water would feed on the sugar and reproduce quickly, using up all the oxygen in the water. Other organisms in the water that need oxygen (like fish) die.
2) Yeast can be used to treat the contaminated water before it's released — it uses up the sugar in respiration.

At yeast it's an easy page...

Yeast releases more energy from aerobic respiration than from anaerobic respiration. This means that when there is a good oxygen supply, the yeast has more energy and so reproduces more.

Brewing

There's more to yeast than cleaning up sugar spills...

We Use Yeast for Brewing Beer and Wine

1 Firstly you need to get the <u>sugar out</u> of the barley or grapes:

BEER

1) Beer is made from <u>grain</u> — usually <u>barley</u>.
2) The barley grains are allowed to <u>germinate</u> for a few days, during which the <u>starch</u> in the grains is broken down into <u>sugar</u> by <u>enzymes</u>. Then the grains are <u>dried</u> in a kiln. This process is called <u>malting</u>.
3) The malted grain is <u>mashed up</u> and water is added to produce a <u>sugary solution</u> with lots of bits in it. This is then sieved to remove the bits.
4) <u>Hops</u> are added to the mixture to give the beer its <u>bitter flavour</u>.

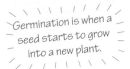

WINE

The grapes are <u>mashed</u> and water is added... a bit simpler than beer making.

Germination is when a seed starts to grow into a new plant.

2
- <u>Yeast</u> is <u>added</u> and the mixture is <u>incubated</u> (heated up). The yeast <u>ferments</u> the <u>sugar</u> into <u>alcohol</u>.
- The fermenting vessels are designed to stop <u>unwanted micro-organisms</u> and <u>air getting in</u>.

> 1) The <u>rising concentration of alcohol (ethanol)</u> in the fermentation mixture due to <u>anaerobic respiration</u> eventually starts to <u>kill</u> the yeast. As the yeast dies, fermentation <u>slows</u> down.
> 2) Different species of yeast can <u>tolerate different levels of alcohol</u>. Some species can be used to produce strong wine and beer with a <u>high concentration</u> of alcohol.

3
- The beer and wine produced is <u>drawn off</u> through a tap.
- Sometimes chemicals called <u>clarifying agents</u> are added to <u>remove particles</u> and make it <u>clearer</u>.

4
- The <u>beer</u> is then <u>pasteurised</u> — <u>heated</u> to <u>kill any yeast</u> left in the beer and completely stop fermentation. Wine isn't pasteurised — any yeast left in the wine carry on slowly fermenting the sugar. This <u>improves the taste</u> of the wine. Beer also tastes better if it's unpasteurised and aged in the <u>right conditions</u>. But big breweries pasteurise it because there's a <u>risk</u> unpasteurised beer will <u>spoil</u> if it's not stored in the right conditions after it's sold.
- Finally the <u>beer</u> is <u>casked</u> and the <u>wine</u> is <u>bottled</u> ready for sale.

Distillation Increases the Alcohol Concentration

1) Sometimes the products of fermentation are <u>distilled</u> to <u>increase</u> the <u>alcohol content</u>. This produces <u>spirits</u>, e.g. if <u>cane sugar</u> is fermented and then distilled, you get <u>rum</u>. <u>Fermented malted barley</u> is distilled to make <u>whisky</u>, and <u>fermented potatoes</u> are distilled to make <u>vodka</u>.

2) Distillation is used to <u>separate</u> the alcohol out of the alcohol-water solution that's produced by fermentation.

3) The fermentation products are <u>heated to 78 °C</u>, the temperature at which the alcohol (but not the water) boils and turns into vapour.

4) The <u>alcohol vapour rises</u> and travels through a cooled tube which causes it to <u>condense</u> back into <u>liquid alcohol</u> and run down the tube into a <u>collecting vessel</u>.

5) Alcohol can only be <u>distilled</u> on <u>licensed premises</u> — you're not allowed to do it in your garden shed.

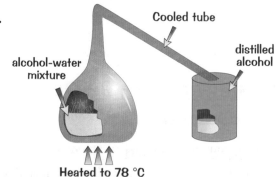

Cooled tube

distilled alcohol

alcohol-water mixture

Heated to 78 °C

Don't try this one at home kids...

You can ferment pretty much <u>any kind of fruit</u> to make alcohol, e.g. cider is made from fermented apples. The <u>sugar</u> that yeast feeds on is naturally found in the fruit — you just have to <u>mash them</u> to get it out.

Fuels from Microorganisms

Food and booze aren't the only things microorganisms can be used for — the stuff they produce can also be used as fuel. And with the world's oil and gas supplies running low, other fuel sources, such as this, are going to become really important.

Fuels Can Be Made by Fermentation

1) Fuels can be made by <u>fermentation</u> of natural products — luckily enough, <u>waste</u> products can often be used.

2) Fermentation is when <u>bacteria</u> or <u>yeast</u> break sugars down by <u>anaerobic</u> respiration.

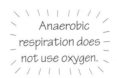
Anaerobic respiration does not use oxygen.

Ethanol is Made by Anaerobic Fermentation of Sugar

1) Yeast make <u>ethanol</u> when they break down <u>glucose</u> by <u>anaerobic respiration</u>.

2) <u>Sugar cane juices</u> can be used, or glucose can be derived from <u>maize starch</u> by the action of carbohydrase (an enzyme).

3) The ethanol is <u>distilled</u> to separate it from the yeast and remaining glucose before it's used.

4) In some countries, e.g. Brazil, <u>cars</u> are adapted to <u>run on a mixture of ethanol and petrol</u> — this is known as '<u>gasohol</u>'.

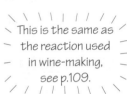
This is the same as the reaction used in wine-making, see p.109.

Biogas is Made by Anaerobic Fermentation of Waste Material

1) Biogas is usually about 70% <u>methane</u> (CH_4) and 30% <u>carbon dioxide</u> (CO_2).

2) Lots of <u>different microorganisms</u> are used to produce biogas. They ferment <u>plant and animal waste</u>, which contains <u>carbohydrates</u>. <u>Sludge waste</u> from, e.g. <u>sewage works</u> or <u>sugar factories</u>, is used to make biogas on a large scale.

3) It's made in a simple fermenter called a <u>digester</u> or <u>generator</u> (see the next page).

4) Biogas generators need to be kept at a <u>constant temperature</u> to keep the microorganisms <u>respiring</u> away.

5) There are two types of biogas generators — <u>batch generators</u> and <u>continuous generators</u>. These are explained on the next page.

6) Biogas <u>can't be stored as a liquid</u> (it needs too high a pressure), so it has to be <u>used straight away</u> — for <u>heating</u>, <u>cooking</u>, <u>lighting</u>, or to <u>power a turbine</u> to <u>generate electricity</u>.

Fuel Production Can Happen on a Large or Small Scale

1) <u>Large-scale</u> biogas generators are now being set up in a number of countries. Also, in some countries, <u>small biogas generators</u> are used to make enough gas for a <u>village</u> or a <u>family</u> to use in their <u>cooking stoves</u> and for <u>heating</u> and <u>lighting</u>.

2) <u>Human waste</u>, waste from <u>keeping pigs</u>, and <u>food waste</u> (e.g. kitchen scraps) can be <u>digested</u> by <u>bacteria</u> to produce biogas.

3) By-products are used to <u>fertilise</u> crops and gardens.

Anaerobics lesson — keep fit for bacteria...

Fascinating stuff, this biogas. It makes a <u>lot of sense</u>, I suppose, to get energy from rubbish, sewage and pig poop instead of leaving it all to rot naturally — which would mean all that lovely <u>methane</u> just wafting away into the atmosphere. Remember — <u>anaerobic respiration</u> makes biofuels.

Fuels from Microorganisms

Here's more than you could ever have wanted to know about that magic stuff, biogas.

Not All Biogas Generators Are the Same

There are two main types of biogas generator — <u>batch generators</u> and <u>continuous generators</u>.

<u>Batch generators</u> make biogas in <u>small batches</u>. They're <u>manually loaded up with waste</u>, which is left to digest, and the by-products are cleared away at the end of each session.

<u>Continuous generators</u> make biogas <u>all the time</u>. Waste is <u>continuously fed in</u>, and biogas is produced at a <u>steady rate</u>. Continuous generators are more suited to <u>large-scale</u> biogas projects.

The diagram on the right shows a <u>simple biogas generator</u>.

Whether it's a continuous or batch generator, it needs to have the following:

1) an inlet for <u>waste material</u> to be put in

2) an outlet for the <u>digested material</u> to be removed through

3) an outlet so that the <u>biogas</u> can be piped to where it is needed

Inlet for waste material

Biogas outlet

Gas

Waste material

Outlet for digested material (to be used as fertiliser)

Four Factors to Consider When Designing a Generator:

When biogas generators are being designed, the following factors need to be considered:

COST: Continuous generators are <u>more expensive</u> than batch ones, because waste has to be <u>mechanically pumped in</u> and digested material <u>mechanically removed</u> all the time.

CONVENIENCE: Batch generators are less convenient because they have to be <u>continually loaded</u>, <u>emptied and cleaned</u>.

EFFICIENCY: Gas is produced most quickly at about <u>35 °C</u>. If the temperature falls below this the gas production will be <u>slower</u>. Generators in some areas will need to be <u>insulated</u> or kept warm, e.g. by <u>solar heaters</u>. The generator shouldn't have any <u>leaks</u> or gas will be lost.

POSITION: The waste will <u>smell</u> during delivery, so generators should be sited <u>away from homes</u>. The generator is also best located fairly close to the <u>waste source</u>.

Using Biofuels Has Economic and Environmental Effects

1) Biofuels are a '<u>greener</u>' alternative to fossil fuels. The <u>carbon dioxide</u> released into the atmosphere was taken in by <u>plants</u> which lived recently, so they're '<u>carbon neutral</u>'.

2) The use of biofuels <u>doesn't</u> produce significant amounts of sulphur dioxide or nitrogen oxides, which cause <u>acid rain</u>.

3) <u>Methane</u> is a <u>greenhouse gas</u> and is one of those responsible for <u>global warming</u>. It's given off from <u>untreated waste</u>, which may be kept in farmyards or spread on agricultural land as fertiliser. Burning it as biogas means it's <u>not</u> released into the atmosphere.

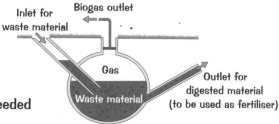

CO_2 released

animal waste

Methane changed to CO_2

CO_2 absorbed in photosynthesis

Biogas generator

4) The raw material is <u>cheap</u> and <u>readily available</u>.

5) The digested material is a better <u>fertiliser</u> than undigested dung — so people can grow <u>more crops</u>.

6) In some developing rural communities <u>women</u> have to spend hours each day <u>collecting wood for fuel</u>. Biogas saves them this drudgery.

7) Biogas generators act as a <u>waste disposal system</u>, getting rid of human and animal waste that'd otherwise lie around, causing <u>disease</u> and <u>polluting water supplies</u>.

Don't sit under a cow — unless you want a pat on the head...

Biogas is <u>fantastic</u>. It gets rid of <u>waste</u>, makes a great <u>fertiliser</u> AND provides <u>energy</u>. Biogas <u>isn't new</u> though — before electricity, it was drawn from London's sewer pipes and burned in the street lights.

Enzymes in Action

Enzymes are molecules made of <u>protein</u>, which <u>speed up (catalyse) chemical reactions</u> in living organisms. Scientists know a good thing when they see it, and enzymes are now used for all sorts of stuff...

Enzymes are Used in Biological Washing Powder...

1) Some stains are caused by <u>soluble</u> chemicals and so they <u>wash out</u> easily in water. Stubborn stains contain <u>insoluble chemicals</u> like starch, proteins and fats. They don't wash out with just water.

2) <u>Non-biological washing powders</u> (detergents) contain <u>chemicals</u> that break up <u>stains</u> on your clothes.

3) <u>Biological washing powders</u> contain the same chemicals as non-biological ones, but also contain a mixture of <u>enzymes</u> which break down the stubborn stains.

Stain	Sources of stain	Enzymes	Product
Carbohydrate	Jam, chocolate	Amylases	Simple sugars
Lipid (fats)	Butter, oil	Lipases	Fatty acids and glycerol
Protein	Blood, grass	Proteases	Amino acids

The <u>products</u> of the enzyme-controlled reactions are <u>soluble in water</u> and so can be easily washed out of the material.

4) Biological washing powders need a <u>cooler wash temperature</u> than non-biological powders because the enzymes are <u>denatured</u> (destroyed) by <u>high temperatures</u> (see p.55). However, some newer powders contain enzymes that are <u>more resistant</u> to heat and so can be used with a hotter water temperature.

5) The enzymes <u>work best</u> at <u>pH 7</u> (neutral). Tap water is usually about pH 7, but in areas with very hard water (water containing high levels of calcium) it might be alkaline, which can damage the enzymes.

6) You can buy <u>special stain removers</u> (e.g. for wine, blood or oil). Some of these are just special solvents, but some contain <u>specific enzymes</u> that will break down the stain.

...and in Medical Products...

1) <u>Diabetes</u> (see p.13) is <u>diagnosed</u> by the presence of <u>sugar</u> in the <u>urine</u>. Many years ago, doctors actually used to taste patients' urine to test for sugar... yuk. Later they tested the urine for sugar using <u>Benedict's solution</u>. When it's heated, the solution <u>changes colour</u> from blue to orange if sugar is <u>present</u>. This test relies on chemical properties (not enzymes).

2) Nowadays, <u>reagent strips</u> (strips of paper with enzymes and chemicals in them) are used. They're dipped in urine and <u>change colour</u> if sugar is <u>present</u>.

3) This test is based on a sequence of <u>enzyme reactions</u>. The product of the enzyme-controlled reactions causes a chemical embedded in the strip to change colour.

4) There are similar strips which can be used to test <u>blood sugar levels</u> (see next page).

...and in the Food Industry

Low-Calorie Food	1) <u>Table sugar (sucrose)</u> is what you normally <u>sweeten food with</u> at home.
	2) In the food industry an enzyme called <u>invertase</u> is used to <u>break down sucrose</u> into <u>glucose</u> and <u>fructose</u>. Glucose and fructose are <u>much sweeter</u> than sucrose.
	3) This means you can get the same level of sweetness using <u>less sugar</u>. This helps to make <u>low-calorie food sweeter</u> without adding calories.
Cheese	The enzyme <u>chymosin</u> is used to <u>clot milk</u> in the first stages of <u>cheese production</u>.
Juice Extraction	The enzyme <u>pectinase</u> is used in <u>fruit juice extraction</u>. It breaks down <u>pectin</u> (a part of the cell wall in apples and oranges), causing the cell to release its juice.

Stubborn stains — not just dirty, but grumpy...

Not everyone can use biological washing powders. Some of the enzymes remain on the clothes and can irritate sensitive skin, making it sore and itchy. Sensitive people have to use non-biological powders.

Enzymes in Action

When enzymes are used to speed up reactions, they end up <u>dissolved in the mixture</u> with the substrates and products — and can be <u>difficult to remove</u>. One way to avoid this is to <u>immobilise</u> the enzymes...

Immobilising Enzymes Makes Them Easier to Remove

1) Many industrial processes use <u>immobilised enzymes</u>, which <u>don't</u> need to be <u>separated out</u> from the mixture after the reaction has taken place.

2) Immobilised enzymes are <u>attached</u> to an <u>insoluble material</u>, e.g. <u>fibres</u> (like collagen or cellulose), or <u>silica gel</u>. Or they are encapsulated in <u>alginate beads</u> (alginate is a jel-like substance).

3) The immobilised enzymes are <u>still active</u> and still help speed up reactions.

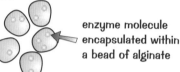

enzyme molecule encapsulated within a bead of alginate

Advantages of Immobilising Enzymes
1) The insoluble material with attached enzymes can be washed and <u>reused</u>.
2) The enzymes <u>don't contaminate</u> the product.
3) Immobilised enzymes are often more <u>stable</u> and less likely to denature at high temperatures or extremes of pH.

Immobilised Enzymes Can be Used to Make Lactose-Free Milk

1) The sugar <u>lactose</u> is naturally found in <u>milk</u> (and yoghurt). It's broken down in your digestive system by the <u>enzyme lactase</u>. This produces <u>glucose</u> and <u>galactose</u>, which are then <u>absorbed</u> into the blood.

2) Some people <u>lack the enzyme lactase</u>. If they drink milk the lactose isn't broken down and gut <u>bacteria</u> feed on it, causing <u>abdominal pain</u>, <u>wind</u> and <u>diarrhoea</u> — these people are <u>lactose intolerant</u>.

3) <u>Cats</u> are also lactose intolerant (which is odd considering how much they like milk). They can't digest lactose and have the same symptoms as humans if they drink it.

4) <u>Lactose-free milk</u> can be produced using <u>immobilised lactase</u>. (Special <u>lactose-free cats' milk</u> is also produced using lactase.)

5) A method called <u>continuous flow processing</u> is often used for this:
- The substrate solution (milk) is run through a <u>column of immobilised enzymes</u>.
- The enzymes convert the substrate (lactose) into the products (glucose and galactose), but only the <u>products</u> emerge from the column. The enzymes stay fixed in the column.

Milk

Column of immobilised lactase

Lactose free milk

Immobilised Enzymes are Also Used in Reagent Strips

1) <u>Diabetics</u> use reagent strips to measure their <u>blood glucose concentration</u> on a <u>daily basis</u>. They're <u>quick</u> and <u>convenient</u> to use. Before reagent strips diabetics had to 'guess' when they needed to inject insulin (e.g. before meals), because there was no quick way of knowing what their glucose level was.

2) There are <u>immobilised enzymes</u> on the reagent strips.

3) A drop of blood from a finger prick is added to the strip. The enzymes in the strip cause it to <u>change different colours</u> depending on the <u>glucose concentration</u>. The colour is then compared to a <u>chart</u> to find out the level of blood sugar.

You treat your cat with milk — he thanks you with diarrhoea...

Lactose intolerance affects <u>millions of people</u>. There's a pretty big industry out there providing them with lactose-free milk, lactose-free ice cream, lactose-free chocolates...

Genetically Modifying Plants

Genetic engineering (another kind of biotechnology) is also on page 43. But you need more details now.

Genetically Modifying Plants is Done Differently from Modifying Bacteria

Genetically modified organisms (GMOs) are made by 'cutting and pasting' genes — you 'cut out' the gene for the characteristic you want, and 'paste it into' the organism you want it in.
In reality, this cutting and pasting takes a bit of doing...

1) To make GM plants, scientists often use a bacterium called Agrobacterium tumefaciens.
   This naturally invades plant cells and inserts its genes into the plant's DNA.

2) If other genes are added to this bacterium, then those genes are taken along too. Works a treat.

3) For example, you could make a herbicide-resistant plant like this...

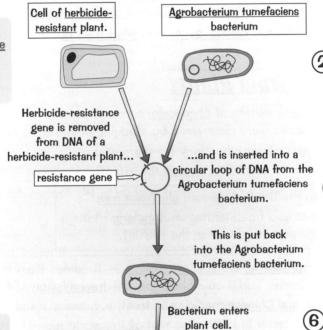

(1) You get a plant that already has resistance to the herbicide, and you work out which gene is responsible.

Cell of herbicide-resistant plant.

Agrobacterium tumefaciens bacterium

(2) You cut out this gene for herbicide-resistance from one of the plant's cells.

Herbicide-resistance gene is removed from DNA of a herbicide-resistant plant...

resistance gene

...and is inserted into a circular loop of DNA from the Agrobacterium tumefaciens bacterium.

(3) Agrobacterium tumefaciens bacteria contain a circular loop of DNA — you remove this from a bacterium, 'cut it open', and insert the herbicide-resistance gene.

This is put back into the Agrobacterium tumefaciens bacterium.

(4) The next step is to allow the genetically modified Agrobacterium tumefaciens bacteria to infect cells of the 'target' plant.

(5) The bacteria will insert their genes (including the herbicide-resistance gene) into the plant's DNA.

Bacterium enters plant cell.

Cell of GM plant.

(6) These cells are then grown on a medium containing the herbicide. Those that grow must contain the herbicide-resistant gene.
Job's a good 'un.

GM Crops May Have Disadvantages

There are lots of 'issues' around GM crops.

1) A big concern is that transplanted genes may get out into the natural environment. For example, the herbicide-resistance gene may be picked up by weeds, creating new 'superweeds'.

2) Some people say that growing GM crops will affect the number of weeds and flowers (and therefore wildlife) that usually lives in and around the crops — reducing farmland biodiversity. (Herbicides will also do this though.)

3) Not everyone is convinced that GM crops are safe. People are worried they may develop allergies to the food — although no unexpected effects have been found in food currently on sale.

Using GM bacteria to make GM plants... genius...

Genetically engineer a bacterium, then use that to genetically engineer a plant — cunning. GM crops are controversial, but if they're as good as some people say, they may help with all sorts of problems.

Developing New Treatments

Biotechnology can also be used when developing new drugs — talk about a finger in every pie. In the future studying genes could also be really important in developing new treatments.

Studying Genes Can Lead to Development of New Medicines

The study of all the genes in an organism is called genomics. Genomics could have useful applications in various fields (e.g. medicine, agriculture...). It's the applications to medicine you need to know about:

1) Determining which genes predispose people to diseases (make people more likely to get them) could lead to better early diagnosis, or even prevention of diseases.

2) Identifying defective genes and what they do can help scientists understand how a disease is caused, and so how to treat it. This might be with a drug or with gene therapy (see p.40).

3) Not everyone responds to the same drug in the same way because everyone has different genes. It may be possible in the future to tailor drugs to an individual's genes.

4) The more we know about the pathogens' genes and what they do, the more likely we are to be able to figure out new ways to kill them.

Many Drugs Come from Plants

1) Plants and animals produce a variety of chemicals, some of which can be used as drugs. A lot of our current medicines were discovered by studying plants used in traditional cures.

2) Scientists identify promising plants, then look for the active ingredient (the chemical in the plant that causes the effect).

Scientists also 'screen' large numbers of plants to see if they contain anything useful.

QUININE
- Quinine comes from the South American cinchona tree.
- For years, it was the main treatment against malaria (though more effective drugs were produced in the 1930s).

Malaria's an infectious disease carried by mosquitoes. It kills over a million people a year.

ARTEMISININ
- One of the newer anti-malarial drugs is artemisinin — it comes from the plant Artemisia annua. It's an effective treatment, and it could help prevent transmission of the disease.
- It was used in traditional Chinese medicine to treat, e.g. malaria and skin diseases.
- Chinese scientists isolated it in 1972. The rest of the world found out about it a bit later.
- Other, more powerful drugs have since been derived from artemisinin.

Drug Development Costs a Lot of Money

1) Modern drugs can be very expensive to produce. When drug companies produce a new drug, they take out a patent on it — this means only they can make it and sell it for a certain number of years.

2) This allows the company to make back the huge amount of money needed to research and develop the drug (p20). But some people say the patent means this company can basically charge what they like.

3) This is controversial — not everyone who needs them can afford treatments still under patent.

Some people think...
- ...drug companies could charge less for their drugs and still make a profit.
- ...making any profit from people's illnesses is unethical.
- ...other companies should be allowed to copy the drugs and reproduce them cheaply.

Some people think...
- ...it'd be unfair if other companies could make and sell the treatment more cheaply — they haven't had to do any of the expensive research (and profit from successful drugs has to cover the costs of research on drugs that are rejected during trials).
- ...there'd be no reason for a company to develop new treatments if they can't make a profit.

I doubt you'd discover medicines if you studied my jeans...

You might not have thought it but even making new drugs can be a bit of a controversial issue. Make sure you know both sides of the argument, they could just sneak a question about it in the exam.

Revision Summary for Section Eleven

I bet you thought you'd never get to the end of this book... well, you're not quite there yet, there's still a whole other section to come. I never knew how interesting microorganisms could be, they do so many things — making yoghurt, causing disease, clearing up sugar spills, making booze and biogas. And as for those useful enzymes... they're a barrel of laughs.

1) State the function of the following parts of a bacterial cell: a) flagellum, b) cell wall, c) bacterial DNA.

2) How do bacteria reproduce?

3) Describe the four stages in an infectious disease.

4) Suggest two reasons why the incidence of cholera and dysentery might be high in a given population.

5) Explain why natural disasters often cause rapid spread of disease.

6) Describe the process of making yoghurt.

7) Name the three types of microorganism used to make soy sauce.

8) Why are plant stanol esters added to spreads?

9) What are prebiotics? Why might some people choose to take them as a supplement?

10) How is chymosin traditionally obtained?
    Where does the chymosin used to make vegetarian cheese come from?

11) What is vitamin C used for? What species of bacterium helps to produce it?

12) How is MSG made? What is it used for?

13) What is carrageenan used for and where does it come from?

14) State the word equations for anaerobic respiration and aerobic respiration in yeast.

15) How is the rate of breakdown of sugar by yeast affected by temperature? Sketch a graph to illustrate your answer.

16) Describe the main stages in brewing beer.

17) How could you increase the alcohol concentration of a fermented product?

18) What are the two main components of biogas?

19)* Loompah is a small village. It's very hot in the summer but freezing cold in winter. The villagers keep goats and cows. They also try to grow crops, but the soil isn't very fertile, so it's difficult. The villagers currently rely on wood for fuel for heating and cooking. There's not much of this around, so they spend a lot of time collecting it, preventing them from practising their nail-art.

    a) How suitable do you think biogas would be for this village? Explain the advantages that using biogas would have for the village. What disadvantages or problems might there be?

    b) Loompah starts using biogas and uses the digested material as fertiliser. They compare their crops to the crops grown by the village of Moompah, which uses normal manure as a fertiliser. Loompah's crops are bigger, so they conclude that the digested material is a better fertiliser than manure. What do you think of the conclusion they've drawn?

20) Which enzyme in biological powder would break down a stain made of: a) butter, b) grass?

21) Why do biological washing powders need a cool wash temperature and neutral pH?

22) Name an enzyme that breaks down sucrose. What is this enzyme used for in the food industry?

23) Give three advantages of immobilising enzymes.

24) Describe how Agrobacterium tumefaciens is used to genetically engineer plants, using the example of herbicide-resistant genes.

25) Describe three possible disadvantages of GM crops.

26) How might the study of genomics help develop new medicines?

27) Describe one way that a drug from a plant might be discovered.

28) Describe the arguments for and against pharmaceutical companies charging high prices for drugs.

29) Why do some people think that drug company profits aren't fair?

* Answers on page 140

Instinctive and Learned Behaviour

Behaviour is a pretty complicated topic, but pretty interesting too...

Behaviour is an Organism's Response to Changes in Its Environment

1) Behaviour is how an organism responds to things going on in its environment — helping it to survive.

2) Behaviour can either be inherited or learned, but most behaviour relies on a combination of the two.

3) Both your genes and your environment play a part in influencing your behaviour and it's sometimes hard to decide what is inherited and what is learned. One example is human speech:

Most humans are born with the instinctive ability to speak, but if a child is born deaf, or is brought up in isolation and doesn't hear people speaking, then it won't learn to speak. Humans have to learn language.

Some Behaviour is Inherited...

This inherited behaviour is known as instinctive behaviour. Animals can respond in the right way to a stimulus straight away, even though they've never done it before, e.g. newborn mammals have an instinct to suckle from their mothers. Instinctive behaviour can be a fairly simple reflex, or a complicated behaviour, like a courtship ritual.

A stimulus is a detectable change in the environment.

Reflex actions are simple inherited behaviours, where a stimulus produces a fairly simple response, like sneezing, salivation, coughing, and blinking. They often protect us from dangerous stimuli. Reflexes are automatic actions — you don't have to think about them.

Some types of reflex are slightly more complex:

1) Earthworms show what's known as 'negative phototaxis' — they move away from light.

2) Sea anemones wave their tentacles more when stimulated by chemicals emitted by their prey.

...and Some Behaviour is Learned

Learned behaviour isn't inherited — you have to learn it, obviously. It lets animals respond to changing conditions. Animals can learn from their previous experiences how to avoid predators and harmful food, and how to find food or a suitable mate.

Habituation

If you keep on giving an animal a stimulus that isn't beneficial or harmful to it, it quickly learns not to respond to it. This is called habituation. This is why crows eventually learn to ignore scarecrows (because they don't harm the bird or reward it). This is also why you can often sleep through loud and familiar noises, like traffic, but might wake up instantly at a quiet but unfamiliar noise. By ignoring non-threatening and non-rewarding stimuli, animals can spend their time and energy more efficiently. This is an especially important learning process in young animals — they are born with an inherited tendency to be frightened by loud, bright, sudden stimuli and they must quickly learn which stimuli to ignore so they can concentrate on stimuli that are possibly dangerous.

Experiences in very early life, when the brain is actively developing, can massively affect later behaviour. Examples include:

1) Pigs removed from their mothers at an early stage become more aggressive adults.

2) Some birds never learn the proper bird song for their species if kept in isolation when they are young.

3) Babies whose parents argue a lot are more likely to suffer from attacks of rage when they get older.

I think I have negative revisiontaxis...

Habituation happens more often than you think — e.g. you learn to ignore the stimuli produced by the weight of your clothes because you're used to wearing them. There is also a type of habituation nicknamed 'banner blindness' where internet users fail to notice advertising banners after a while.

Instinctive and Learned Behaviour

Habituation isn't the only type of learned behaviour — there are more...

Conditioning is Another Form of Learned Behaviour

There are two types of conditioning:

Classical Conditioning

Classical conditioning happens when an animal learns PASSIVELY (i.e. without actually trying) to associate a 'neutral stimulus' with an important one, e.g. a dog associates a bell ringing with the arrival of food. The response is automatic and reinforced by repetition.

Example: Ivan Pavlov — Classical Conditioning in Dogs
Pavlov studied the behaviour of dogs and noticed that they would salivate (drool) every time they saw or smelt food. He began to ring a bell just before each time the dogs were given their food. After a while he found that the dogs salivated when the bell was rung even if he didn't give them their food.

Operant Conditioning

Operant conditioning or 'trial and error learning' is where an animal learns ACTIVELY to associate an action with a reward or a punishment. (So the animal actually tries to work out what's going on.) This happens in humans when children are rewarded or punished for specific behaviour.

Example: Burrhus Skinner — Operant Behaviour in Pigeons and Rats
Skinner trained rats and pigeons to obtain a food reward using a small cage that he invented (called a 'Skinner box'). The animal had a choice of buttons to press. When the animal pressed a particular lever or button, it was rewarded with food. He found that pigeons and rats used a system of trial and error to learn which button to press to get the reward.

We Use Conditioning to Train Animals

Humans use both classical and operant conditioning to train animals to do certain things.

Training animals usually involves operant conditioning — giving rewards when the animal does what you want or punishments when it does something you don't want it to do. Rewards like food treats work best, but sometimes just praise will do. Punishments can be physical (like choke chains which pull around a dog's neck if it pulls on the lead) or verbal (like saying 'No!'). However, punishment isn't recommended any more for animal training — it's stressful for the animal and rewards work just as well.

Here are some examples of animal training using operant conditioning:

- Training guide dogs to stop at a roadside and wait for a command.
- Training police sniffer dogs to retrieve drugs.
- Training animals to 'act' in films, for example horses falling down as if they have been shot.

Classical conditioning is used in combination with operant conditioning when the reward can't be given at the exact time the act is carried out. For example, a dolphin can't always be rewarded with fish at the exact moment it does a jump. The trainer gets the dolphin to learn to associate a whistle with getting fish, then whistles when the animal does the jump. The whistle is, in a way, the reward, as it tells the dolphin that it will get a fish.

I condition my hair to make it lie down...

Here's how to condition your teacher. If your teacher is one of those active types that like to stroll backwards and forwards while talking to you, simply look interested while they're on one side of room, and bored while they're on the other. In no time, they'll only be teaching from one side of the room.

Social Behaviour and Communication

You thought it was only humans that had social lives and nice chats? Wrong!

Animals Need to Communicate

Communication between different individuals in a group is beneficial in a number of ways:

1) It can help keep the group together.
2) If any one animal sees a predator, it can warn all the others.
3) Communication of mood can avoid unnecessary fighting.
4) Baby animals can communicate their needs to their parents.
5) Communication can allow predators hunting in a pack to coordinate their attack.

Animals Can Communicate in Different Ways

SOUND

Communication by sound is pretty common in nature, and occurs in humans too — through language.

- Whales and dolphins can communicate over long distances using low-frequency sound.
- Birds' calls are used to declare their territory, attract a mate or warn others about predators.

CHEMICALS

Chemicals called pheromones can be released by an animal to tell others where it is or where it has been:

- Many animals use chemical 'scents' to mark the boundaries of their territory, e.g. dogs pee on things.
- Other chemicals can act as sexual attractants. In some moths, the male can detect the female's pheromone even if he's several kilometres away from her.

BEHAVIOUR

Some animals use specific behaviour signals to communicate:

- Honey bees move in a certain way, called a 'waggle dance', when they return to the hive to tell others where they've found food.

- Most mammals can communicate certain intentions through their body posture (how they hold themselves) and gestures (small movements).

- For example, many use behaviours to threaten others — to intimidate them and so avoid an actual fight. Chimps do this by staring or raising an arm.

- Just as behaviours are used to threaten, they're also used to admit defeat — for example, a dog rolling on its back is showing submission.

- There are plenty of courtship behaviours in different species too — from funny dances to offering gifts to building elaborate nests.

Male peacocks raise and shake their fancy tail feathers as a courtship display.

Facial Expressions are Species-Specific

Another sort of communication signal is facial expression. We're all familiar with human expressions, but we shouldn't apply them to other animals. Facial expressions mean different things in different species — they're species-specific.

A chimpanzee that appears to be 'smiling' is actually expressing fear if its teeth are exposed, and threatening you if its lips are closed.

To us, this chimp looks like he's laughing, but really he's showing fear.

Bad dancing — not very attractive to a female of any species...

Spare a thought for all the chimps you've ever seen in TV ads. They'll be dressed up in human clothes, pouring each other tea, and apparently sharing a joke. But in order to make their 'laughter' facial expression, they weren't tickled or told something funny — they were probably frightened. Poor guys.

Social Behaviour and Communication

We humans are great — we can talk, signal, gesture, and we even know who we are and what we're doing.

Humans Have Complex Ways of Communicating

There are loads of ways that you can communicate things to others, whether you mean to or not:

LANGUAGE

Language is the most obvious form of human communication — it can be spoken, written or signed, and is used to transmit knowledge of past events, emotions and complex ideas to other humans. Language is intentional — you only use it when you consciously mean to communicate something. It's also symbolic — words are used to represent objects or ideas. For example, we use the word 'carpet' to represent a soft covering for floors. Such complex language is unique to humans, although it is possible that some great apes have a very simple form of language. The way we speak is important, too — the volume and tone of speech can indicate emotions like excitement, anger, fear etc.

NON-VERBAL COMMUNICATION

Non-verbal communication in humans is also pretty complex. Some types of non-verbal communication are intentional, and others aren't. The intentional, conscious types are different around the world, like languages are. For example, in Britain we indicate 'no' by shaking our heads, but in Greece they jerk their heads backwards. Other types of non-verbal communication are automatic (although they can be suppressed if we try), and are the same all around the world:

Unconscious facial expressions are the same for everyone. We all raise our eyebrows in surprise, smile or laugh when we're happy, cry when we're sad and screw up our faces in disgust. These facial expressions are really useful for telling others how we feel about things.

Body language is also a way in which we can unconsciously communicate our feelings, for example: Pointing your leg or body towards someone may indicate your interest in them. Standing with your hands on your hips may indicate aggression. Showing your open palms may indicate honesty. Avoiding eye contact may indicate shyness or deception. There are loads of other examples.

All these forms of communication go on at the same time. Language is important, but research indicates that most people react far more to non-verbal signals than to what people are actually saying.

Some People Think Humans are More Self-Aware

Some scientists believe that one thing that distinguishes humans from other animals is self-awareness. The problem is, people don't all define self-awareness the same way. Here are two different definitions:

(1) The most basic definition of self-awareness is being aware of your own existence. It's often tested by showing an animal its own image in a mirror. If it thinks it's seeing another animal, it isn't self-aware, but if it realises that what it's seeing is itself, it is self-aware. Human babies can do this at a very early age, but so do some other animals, like chimps and dolphins.

(2) Some people think self-awareness is also about being aware of your own behaviour and feelings (consciousness), and the possible outcomes of your behaviour (accountability). For example, you know that if you watch your favourite programme on the TV, you will enjoy it. You also know that if you eat all your brother's sweets, he'll get angry.

It's hard to tell if animals have consciousness and accountability, because we can only see what they do — we can't tell what they're thinking when they do it. So the level of self-awareness in animals is open to debate, but we're pretty sure that humans are more self-aware than other animals.

Crossed arms is defensive behaviour...

If you've got a baby brother or sister, you can try the self-awareness experiment out at home. Stick a red dot on their forehead, leave them to forget about it, then sit them in front of the mirror. If they start feeling their forehead then it proves that they know their reflection is them and not a different baby.

Feeding Behaviours

Herbivores and carnivores have really different lifestyles — and it's all because of their choice of food.

Feeding Behaviour Depends on What You Eat

If you eat a certain type of <u>food</u>, you'll need to use a certain type of <u>behaviour</u> in order to get your food in the most <u>efficient</u> way. This is why the feeding behaviour of <u>herbivores</u> and <u>carnivores</u> is quite different.

Herbivores

Herbivores are animals that <u>eat plants</u> and not other animals. Rabbits, cows, sheep, deer, horses and guinea pigs are all herbivores. Here are the main points about these guys:

1) Herbivores <u>don't</u> need to <u>catch</u> their food, which is a bonus.

2) However, because vegetation is <u>low</u> in some <u>essential nutrients</u> like amino acids needed for growth, they need to eat a <u>lot</u> of plants to get enough nutrients. Plants are also more <u>difficult to digest</u>, which makes it even harder to get those essential nutrients out.

3) So, herbivores have to spend a lot of <u>time</u> eating in order to get enough nutrients. Donkeys, for instance, spend six to seven hours feeding every day.

Hard work, this eating lark.

4) Spending so long eating can be quite <u>dangerous</u> — it's hard to spot <u>predators</u> when you're eating.

5) <u>Vertebrate</u> herbivores like wildebeest and buffalo often feed in large groups or <u>herds</u> for <u>safety</u>. At any given time, there'll always be some of the group <u>not feeding</u> who may be able to <u>spot</u> predators and warn the herd. Some members will get <u>caught</u>, but it's still safer to be in a group than alone.

6) The problem with feeding in a group is that the herd will quickly <u>eat up</u> all the food in one area, so they often have to <u>travel</u> large distances to get enough food.

As herbivores are the main food of most carnivores, herbivores have <u>evolved</u> methods of <u>spotting</u>, <u>avoiding</u>, <u>fleeing</u> from or <u>resisting</u> predation. <u>Antelopes</u>, for instance, are very <u>quick</u>. <u>Buffaloes</u> have strong <u>horns</u> which they use to defend themselves and the herd. And finally, most herbivores have <u>eyes</u> on the <u>sides</u> of their head rather than the front. This means that they can see almost all <u>around</u> them, making it easier to <u>spot</u> predators.

Carnivores

Carnivores are animals that <u>eat meat</u>, i.e. other animals. Meat is a <u>nutrient-rich</u> food, so they <u>don't</u> need to spend long <u>feeding</u> compared to herbivores. Lions and tigers will eat a <u>large</u> amount of food when they catch prey, but then may go several days <u>without feeding</u> at all.

A born killer.

Successful carnivores are <u>adapted</u> for <u>detecting</u> and <u>catching</u> their prey (<u>predation</u>). They have <u>eyes</u> at the <u>front</u> of their head, which helps them to judge <u>distance</u> when catching prey. They are usually <u>quick</u> and <u>powerfully built</u>, with <u>sharp teeth</u> and <u>claws</u> to help them kill.

Some carnivores hunt in <u>packs</u> and others hunt on their <u>own</u>. Hunting in packs is useful for rounding up and catching <u>large animals</u>, for example <u>lions</u> hunt in packs to catch <u>zebra</u> and <u>water buffalo</u>. Because the prey is large, it will be enough to feed <u>several animals</u>. Carnivores that eat <u>smaller</u> animals tend to hunt <u>individually</u> — what they catch only has to feed themselves.

For example, <u>foxes</u> hunt individually for small animals like <u>rabbits</u>. <u>Wolves</u> hunt in <u>packs</u> for <u>large</u> animals like deer, but also hunt <u>individually</u> for <u>smaller</u> prey such as squirrels and mice.

Imagine having to eat for hours a day...

Remember, these diets are what the animals <u>primarily</u> eat. Herbivores might eat bones if they have a <u>nutritional deficiency</u>, and carnivores sometimes eat plants to help them <u>vomit</u>. Carnivores won't turn down <u>sweet</u> or <u>milky</u> food if they're offered them either. And don't forget <u>omnivores</u> — they eat plants <u>and</u> meat.

Feeding Behaviours

Now that we've looked at <u>what</u> animals eat, it's time to consider <u>how</u> they do it. Knife and fork perhaps?

Mammals and Birds Feed Their Young

Mammals and birds show <u>parental care</u> — they look after and <u>feed</u> their young
for some time after birth. Mammals feed their young on <u>milk</u>, whereas birds
bring back food to the nest and then <u>regurgitate</u> it to feed their chicks.

Feeding the babies is
pretty easy for mammals.

Certain behaviours will <u>indicate</u> to the mother that her young <u>want feeding</u>:

1) In mammals, this is usually the young <u>sucking</u> on the mother's teat —
   milk will be automatically <u>released</u> to feed the baby.

2) In birds, it's a bit more complex. The young birds may <u>call</u> to their parents in order
   to be fed, or open their <u>mouths</u> wide, showing off the <u>bright colour</u> inside their mouths.
   These <u>stimuli</u> cause the parents to <u>regurgitate</u> food into the mouths of their chicks.

> Young <u>herring gulls</u> peck on a <u>red spot</u> on the mother's beak, and
> this <u>stimulates</u> the mother to <u>regurgitate</u> food. The young birds are
> '<u>programmed</u>' to do this when they're hungry (i.e. it's instinctive
> behaviour). They will peck at <u>any</u> red spot when they're hungry —
> it doesn't have to be a real beak. Even a <u>red rubber</u> on the end of
> a pencil will be pecked at by hungry herring gull chicks.

Red spot

Some Animals Use Tools to Get Food

It used to be thought that <u>humans</u> were the only animals that used tools, but in recent years <u>monkeys</u>,
<u>apes</u> and <u>birds</u> have also been seen to use tools. Here are four examples of animals which use tools:

> <u>Chimps</u> build tools from <u>twigs</u>, which they use to get <u>ants</u>
> out of their holes. They also use sticks to get <u>honey</u> from
> beehives, and to dig up edible <u>roots</u>. Chimps also use
> <u>leaves</u> to wipe dirt, blood, and fruit from their fur.

Gertrude was becoming quite
sophisticated in her use of tools.

> The <u>woodpecker finch</u> uses a <u>spine</u> from a <u>cactus</u> to lever
> <u>grubs</u> out from tree bark. The finch uses the spine in its
> beak, then holds it under its foot while eating the grub.
> It then carries the spine to the next branch and reuses it.

> <u>Hooded monkeys</u> in a lab were given a situation where yoghurt was available
> in narrow plastic tubes fixed to the table. The monkeys managed to get the
> yoghurt out by making '<u>spoons</u>' from pieces of wood which were available.

> <u>Egyptian vultures</u> eat ostrich eggs, which have shells that are too hard to break open by pecking.
> The vultures have learned to throw <u>rocks</u> at them to break them instead. Other species of birds
> break eggs by throwing the eggs at <u>rocks</u>. This behaviour isn't considered to be tool use though,
> because to be classed as tool use, the rock must be used as an <u>extension</u> of the bird's own <u>body</u>.

I think I'll stick to cutlery...

You find different sorts of tool use in different <u>populations</u> of the same species. For example, some
chimps might use an 'ant stick', and leaves as wet-wipes — but some other chimps fifty miles away will
use a 'honey stick', and leaves as umbrellas. This proves that such behaviours are <u>learned</u>, not genetic.

Reproductive Behaviours

Ooh... mating and reproduction. No need to get all shy though, it's only about moths and lions doing it.

To Reproduce Sexually You Need to Find and Select a Mate

Finding a potential mate is fairly easy if an animal lives in a social group, but many animals live in isolation and only spend time with others during the mating season. They have to behave in a way that will allow them to find a mate. Here are a few facts about finding a mate:

1) A lot of animals make some sort of song or call to attract a mate, as in many birds, whales and frogs. It is usually the males who make the call, to attract females to them.

2) Some insects use chemicals called pheromones as sexual attractants, but here it's usually the female who produces the signal. In moths, the pheromone can be detected by a male several kilometres away, and he can follow the trail to find the female (see p.120).

3) Sometimes, males actually fight each other and only the winners get to mate, e.g. deer do this. Obviously, it wouldn't be good for the species if lots of males got injured or even killed, so instead of real fighting, the fights often involve displays. These displays indicate strength and give the weaker male a chance to back away.

4) Courtship displays usually involve the male doing a special display to impress the female. They involve things like exaggerated posturing, dancing and showing brightly coloured parts of the anatomy. Courtship displays are species-specific — so the female knows she is mating with a male of the right species. There is often a link between the impressiveness of the display and the fertility of the male. In the mandrill, a mammal with a very brightly coloured face and bum, the brightness of the colours is linked with the level of testosterone, the male sex hormone.

Because the female of most species puts so much more effort into child-rearing than the male, it's important that she doesn't mate with a male of a closely related species. If she did, she'd produce infertile offspring and all her efforts to pass on her genes would be wasted. It's also best if the male is strong and fertile, as this ensures that the next generation will have the best possible chance of survival.

This is why females tend to select a mate, and males have to show that they're worthy of selection. It's not so important for males to be choosy — their job is simply to mate with as many females as they can.

The mandrill: look at his silly nose.
(Shame you can't see his bum.)

Most Animals Have More Than One Mate

Monogamy (staying with just one mate) occurs mostly in birds. It's pretty rare in the rest of the animal kingdom. Most animals have more than one mate, but mating patterns vary between species:

1) In most species, the male takes no part in the birth or care of the young, so there's no reason for him to stick around. Instead he'll go off and mate with other females during the same mating season.

2) In some species (some birds, for instance), he'll mate with one female each season, though not necessarily the same one from year to year.

3) In some mammals (e.g. the lion), a male may have a group of females which he stays with, but mates with all of them. These females are known as his 'harem'.

4) The few animals that are monogamous include the following:
   • Birds — albatross, bald eagle, swan, mallard, raven, penguins and parrots
   • Mammals — gibbons and prairie voles

Male sea lion with his harem

Showing your bum will only attract the police...

So then... what about humans? Some cultures are not monogamous. In these it's fine for men to have more than one wife. Other cultures have a kind of 'enforced monogamy', where society or religion say that you should have one partner only. In these cultures you find that there's a lot of secret infidelity.

Reproductive Behaviours

Our parents often look after us for our first eighteen years — that's crazy compared to most animals.

Some Animals Look After Their Young

Most animals give birth to their young and then leave them to fend for themselves. If they lay eggs, they may incubate and protect the eggs until they are hatched, and then leave. However, in some species, one or both parents look after the young in a variety of ways for different lengths of time. The care may involve protecting them, feeding them and teaching them basic skills. This level of care is mostly seen in birds and mammals, although crocodiles and some fish also care for their young.

PROTECTION

Protection may just involve one parent staying with the young to keep them together and to fend off predators. In some cases, protection is helped by the construction of elaborate nests to enclose the young. The weaver bird weaves strands of leaves and twigs into a ball, sometimes with a long tube attached, which makes it difficult for predators to take, or even notice, the young.

FEEDING

Some feeding behaviours have been covered on pages 122-123 (so have a look). If a species both feeds and protects its young, this usually means that both parents need to be involved — one to stay with the young, the other to get food.

TEACHING SKILLS

Certain behaviours are instinctive and baby animals will learn them without being taught — e.g. walking in mammals, and flying in birds. Other skills need to be taught. Birds called oystercatchers get food by opening mussels, a difficult task which an experienced bird can do in less than a minute, but one that takes months for the young to learn. Human babies need to be taught a whole range of skills from how to get dressed to eating with a knife and fork (or chopsticks). Apart from in humans, where language is very useful in teaching, babies usually learn by simply imitating their parents' behaviour.

Looking After Young Increases Their Survival

Looking after the young puts the mother (in particular) at risk. Food has to be shared, and a lot of time has to be spent with the eggs and baby animals. If the parents protect the young from predators, they decrease their own chances of escaping. Here's why they bother:

1) Parental care greatly increases the proportion of the young that survive. In birds that care for their eggs and young, about 25% of the eggs will produce adult birds. This is high compared to most animals — for example, fewer than one in a million cod eggs survive to become adult fish.

2) Looking after young is less risky for the mother than being pregnant, which puts strain on her body and makes it more difficult to escape predators. If animals care for their young, it means they can give birth to a less developed baby, and so have a shorter pregnancy and spend less time at risk. This only applies to mammals though, as the embryos in birds' eggs are undeveloped when laid.

3) From the species' point of view, it is the survival of an animal's genes into the next generation that is important, rather than the survival of the animal itself. This may be why an animal will risk death (and the loss of one copy of its genes) to protect, for example, four offspring, which all contain its genes.

Go and thank your parents for their efforts right now...

Think about all the things that human parents and carers do — they keep their kids well-fed, clean, warm, happy, healthy and safe, and they keep doing it for years. They most likely hold down a job in order to pay for all the things that kids need too. And all a cod does is lay some eggs and leave. Pah.

Living in Soil

Soil is made up of bits of <u>rock</u>, <u>dead material</u> (like dead leaves and animals), <u>living things</u>, <u>air</u> and <u>water</u>. Soil is teeming with life — insects, bacteria, worms and loads more icky creepy crawlies... eugh.

Soil is Full of Living Things

1) Soil may not look all that exciting, but it's pretty important to us. Plants need it for <u>anchorage</u> (to stop them falling over) and for a <u>supply of minerals</u> and <u>water</u>. And animals need plants for food and oxygen.

2) Soil is an ecosystem in itself, containing complex <u>food webs</u>. <u>Herbivores</u> (plant-eaters), <u>carnivores</u> (meat-eaters) and <u>detritivores</u> (which feed on dead organisms) are all found in the soil.

3) There are several other types of organism that live in the soil — <u>microscopic protozoans</u>, <u>fungi</u>, <u>nematode worms</u> and <u>bacteria</u>.

4) In order for a soil to support life, it must contain <u>water</u> and <u>oxygen</u>. All living things need water to <u>carry out reactions</u> in their cells, and cannot survive without it. Almost everything needs <u>oxygen</u> too, for <u>respiration</u> (see p.60). For example, the roots of plants need to get oxygen from the soil so they can respire.

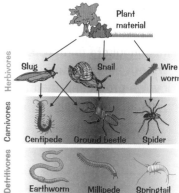

A soil food web

Earthworms Help Keep Soil Healthy and Fertile

<u>Charles Darwin</u>, more famous for his theory of natural selection, spent an awful lot of his time <u>studying worms</u>. He observed them closely and experimented on them to see what sort of food they ate and how they behaved. He discovered these <u>reasons</u> why worms are <u>good for soil</u>:

1) Earthworms <u>bury leaves</u> and other <u>organic material</u> in the soil, where <u>bacteria</u> can <u>decompose them</u>.

2) Their <u>burrows</u> allow <u>air to enter the soil</u> and <u>water to drain through it</u>. Aeration provides the soil organisms with <u>oxygen</u>, but drainage is important, too — if the soil is <u>waterlogged</u>, there is <u>less oxygen</u> available.

3) They <u>mix up</u> the <u>soil layers</u>, <u>distributing</u> the <u>nutrients</u> more <u>equally</u>.

4) Soil in earthworm poo is <u>less acidic</u> than the soil they eat. This can help to <u>neutralise soil acidity</u>, although worms tend to avoid very acidic soils. Acidic soils are <u>less fertile</u> than neutral or alkaline soils.

<u>Farmers</u> and <u>gardeners</u> can <u>buy earthworms</u> (from worm farms) and add them to their soil to improve it.

Bacteria are Involved in Recycling Elements

1) Some elements are very important to living organisms, such as <u>nitrogen</u>, <u>sulfur</u> and <u>phosphorus</u>. Farmers sometimes add <u>fertiliser</u> (containing these elements) to the soil of their fields. But in the natural environment, there is no outside supply of these nutrients, so it's <u>essential</u> that they're <u>recycled</u>. If they weren't, they would <u>run out</u> and the plants would die... not good.

2) <u>Bacteria</u> are important in the recycling of elements, as many of them can change chemicals into other more useable ones. Different bacteria are involved in recycling different elements. E.g. in the <u>nitrogen cycle</u>, the following bacteria play a key part:
   - <u>Saprophytic bacteria</u> in the soil start to <u>decompose dead material</u> into <u>ammonium compounds</u>.
   - <u>Nitrifying bacteria</u>, such as <u>Nitrosomonas</u> and <u>Nitrobacter</u>. Nitrosomonas converts ammonium compounds into <u>nitrite</u>, and nitrobacter converts nitrite into <u>nitrate</u> (which plants can use).
   - <u>Nitrogen-fixing bacteria</u> like <u>Azotobacter</u>, <u>Clostridium</u> and <u>Rhizobium</u>, convert <u>atmospheric nitrogen</u> into useful <u>nitrogen compounds</u>. See p.91 for more on the nitrogen cycle.

Burying leaves — what a fun hobby...

In the exam you might get a question on <u>soil food webs</u>, e.g. 'What happens if you remove centipedes from the soil?' Well, the number of slugs and snails will increase (because fewer are being eaten) and the number of detritivores will go down (because they've got fewer dead centipedes to feed on).

Living in Water

Life in water is very different from life on land. The biggest challenge is regulating water content.

Living in Water Has Its Advantages...

1) One advantage of living in water is that there's a plentiful supply of water... unsurprisingly. There shouldn't be any danger of water shortage or dehydration (unless a drought makes streams dry up).

2) In water, there's less variation in temperature. Water doesn't heat up or cool down as quickly as air, so you don't normally get sudden temperature changes — which water life can find difficult to withstand.

3) Water provides support for plants and for animals that have no skeletal system. E.g. jellyfish are umbrella shaped in water (so they can swim) but if they get washed up on a beach they end up as quivering blobs, because there's not enough support... and then you stand on them. Ouch.

4) Waste disposal is easier. Poo and wee are easily dispersed. The loss of water in wee doesn't matter because there's plenty of water about to make up for it.

...and Its Disadvantages

1) Water is more resistant to movement than air, so animals living in water have to use more energy to move about. Think how much effort it takes to walk in the sea compared to walking on the beach.

2) Aquatic animals have to be able to control the amount of water in their body (water regulation). This is because the water an animal lives in has a different concentration of solutes from the animal's cells. If the animal couldn't regulate water, water molecules would enter or leave the animal's cells by osmosis to even up the solute concentrations. This would cause damage to the cells. E.g.

> • If the animal lived in salt-water its cells would probably contain a lower solute concentration than the surrounding water. If the animal wasn't able to regulate water, then water molecules would leave its cells by osmosis, causing them to shrivel and die.
> • If the animal lived in freshwater, its cells would probably contain a higher solute concentration than the surrounding water. If the animal wasn't able to regulate water, then water molecules would enter its cells by osmosis, causing them to swell and burst.

Water Content is Regulated in Different Ways

1) The kidneys of fish are specially adapted to either salt-water or freshwater to ensure that the concentration of water in the blood remains constant. Some types of fish move between salt-water and freshwater environments and need further adaptations, e.g.

> Salmon live in the sea but move into freshwater rivers to breed.
> Their hormones adjust their bodies to cope with the different environments.

2) Single-celled organisms, like amoebas, only have a cell membrane between them and the surrounding water. They use a different method of water regulation:

> Amoebas regulate water with a contractile vacuole which collects the water that diffuses in by osmosis. The vacuole then moves to the cell membrane and contracts to empty the water outside the cell.

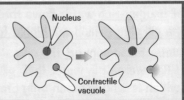

"Waiter, do you have frog's legs?" — "No, I always walk like this"...

Some organisms (mainly insects and amphibians) spend part of their life cycle in water and part on land to exploit both habitats. The two environments provide different challenges, so the different parts of the life cycle often have different body forms (e.g. tadpole and frog).

Living in Water

There are monsters in the water. Millions of them. They're only tiny, mind...

Plankton are Microscopic Organisms That Live in Water

1) Plankton are <u>microscopic</u> organisms that live in <u>fresh</u> and <u>salt water</u>. There are <u>two</u> types:
   - <u>Phytoplankton</u> are microscopic <u>plants</u>.
   - <u>Zooplankton</u> are microscopic <u>animals</u>. Zooplankton <u>feed on</u> phytoplankton.

2) Phytoplankton <u>photosynthesise</u> and are the main <u>producers</u> in <u>aquatic food webs</u>, so they're very important in both freshwater and salt-water ecosystems.

3) Plankton <u>can't move far</u> by themselves and so rely on <u>water currents</u> to carry them from place to place.

4) <u>Phytoplankton</u> populations usually <u>increase</u> between late <u>spring</u> and late <u>summer</u>. This is called an <u>algal bloom</u> (phytoplankton are a type of algae). An algal bloom makes the water go all green and murky.

   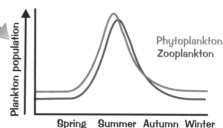

   The increase is due to <u>longer, sunnier days</u> in summer:
   - <u>More light</u> is available for <u>photosynthesis</u> and the energy is used for <u>growth</u>.
   - <u>Temperatures increase</u>, causing both <u>photosynthesis</u> and <u>growth rates</u> to increase.

   The population of <u>zooplankton</u> also <u>increases</u> because there is <u>more phytoplankton</u> to <u>feed on</u>.

5) An increase in <u>nitrates</u> and <u>phosphates</u> also causes algal blooms because the phytoplankton have <u>more nutrients</u>. It happens when water is <u>polluted</u> by <u>fertilisers</u> or <u>sewage</u>.

> You might be asked to <u>interpret marine food webs</u> in the exam. Think about how other organisms in the food chain will be affected by an <u>increase/decrease of plankton</u>. If their food source decreases, so will their population. Keep your eye on what <u>season</u> it is too.

Animals can be Adapted for Gaseous Exchange in Water

In water, animals still need to exchange oxygen and carbon dioxide.
Some animals, such as amphibians and fish have adapted different ways of doing this...

1) <u>Adult</u> amphibians have simple <u>lungs</u>, but their <u>skin</u> also plays an important part in <u>gaseous exchange</u>.

2) <u>Oxygen</u> moves into the animal and <u>carbon dioxide</u> moves out through the <u>skin</u> (as well as via the <u>lungs</u>). To help with this, an adult amphibian's skin has to be kept <u>moist</u>.

3) However, this means the skin can't be <u>waterproof</u>. This lack of waterproofing means the amphibian would <u>lose</u> too much water if it lived in a <u>dry</u> environment.

1) In <u>fish</u>, <u>gas exchange</u> occurs at the <u>gills</u> (slits near the side of the head). A constant supply of <u>oxygen-rich</u> water flows through the open mouth of the fish, and is then forced through the <u>gill slits</u> (which are <u>highly folded</u> to increase the <u>surface area</u>).

2) Water helps <u>support</u> the gills — it keeps the gill folds separated from each other. If fish weren't in water their gills would stick together and they would suffocate (which is why <u>fish</u> can <u>only</u> breathe when they're <u>in water</u>).

What did the fish say when it swam into a wall? Dam...

Different animals do 'gaseous exchange' in different ways, though the <u>basic aim</u> is always the <u>same</u> — to get oxygen <u>into</u> the blood and carbon dioxide <u>out</u>. But all these methods can restrict animals (including us) to a certain kind of <u>habitat</u>. Learn this stuff, or you'll be a fish out of water on exam day.

Human Evolution and Development

We humans have come a long way since our ancestors were swinging about in trees...

Humans and Great Apes Share Common Ancestors

Pygmy chimps (bonobos) are our closest living relatives. Studies of DNA suggest humans and chimps shared a common ancestor that lived in the African rainforest 5-6 million years ago. This ancestral species evolved into two groups — one of which gave rise to modern chimpanzees, the other to humans:

1) The oldest human ancestor is thought to be Australopithecus afarensis (although some people disagree), from 3.5 million years ago. It could walk on two legs, but still spent a lot of its time in the trees.

2) About 2 million years ago, Homo erectus had appeared in Africa, the first species that could be regarded as really human. Homo erectus was a hunter-gatherer living in small family groups. About 1.9 million years ago, they began to leave Africa for the first time and migrate to Asia and then into Europe.

3) Homo erectus evolved into Homo heidelbergensis and lived in Europe about 800 000 years ago. The discovery of spears and other artefacts show that these humans hunted animals. Hunting requires cooperation, so there must have been some social organisation.

4) Meanwhile, Homo erectus survived for a long time in the Far East. 'Java Man' was an example found in Indonesia.

5) About 200 000 years ago, human evolution branched, giving rise to Neanderthal man, or Homo neanderthalensis, in Europe, and Homo sapiens (modern humans) in Africa.

6) Homo sapiens were the more successful species, and when they arrived in Europe about 40 000 years ago, the Neanderthals gradually became extinct, dying out about 28 000 years ago.

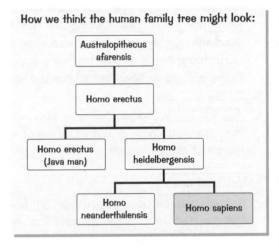

How we think the human family tree might look:

Although we are still the same species, human society has developed considerably since 190 000 years ago. A major factor in this is the use of tools, and in particular being able to create tools for a specific purpose — something which we humans are really good at. Tools enabled us to modify our environment — farming, clearing land and building. This has led to an even more complex society, with people having specific occupations within it.

Humans Developed by Exploiting Animals

Humans have a long history of exploiting other animals:

1) Dogs were domesticated about 14 000 years ago and almost certainly used to help with hunting.

2) As the human population grew after the last ice age, it was difficult to supply all the food needed just by hunting and gathering. Humans developed a farming system with captive herds supplying a ready source of meat. Farming is known to have taken place in some parts of the world 10 000 years ago, although it might have started even earlier, alongside hunting and gathering.

3) Tools allowed us to cultivate land and grow crops. Early farmers also started using animals to actually help them with the farm work, as well as providing meat. For example, there is evidence that humans were riding horses 6000 years ago, and sheep and goats were certainly domesticated in the Middle East 9000 years ago.

Exploiting turkeys.

4) Over the years, selective breeding has resulted in domestic animals that are safer to handle and more productive. For example, cattle have been bred to provide either meat or milk, and to be docile.

All this in just a few thousand years...

Modern farming has come a long way since the old days of hunter-gathering. Having said that though, don't make the mistake of saying that all humans take part in modern farming practices. There are plenty of tribal communities out there who still hunt and gather and don't fancy changing their ways.

Section Twelve — Behaviour in Humans and Other Animals

Human Behaviour Towards Animals

Exploiting animals is a controversial issue, and you need to know both sides of the story.

Humans Exploit Animals

Animals provide <u>meat</u> and <u>farm labour</u>, but humans have also used them in many other ways:

Clothing and domestic materials

Animal hides are used for <u>clothing</u> and <u>upholstery</u>. For example, cow hides (<u>leather</u>) are used to make shoes, bags, jackets and to cover furniture like sofas. The skins from furry animals are sometimes used to make <u>fur coats</u>. <u>Wool</u> from sheep, goats and even rabbits is used to make woolly clothes, carpets and other furnishings.

Entertainment

Horses, dogs and even snails are used for <u>racing</u> — people enjoy <u>betting</u> on which one will win. <u>Bullfighting</u> and <u>cock-fighting</u> are spectator sports in some parts of the world, and again they are used for <u>gambling</u>. Animals also provide entertainment in <u>circuses</u>, <u>zoos</u> and <u>wildlife parks</u>, either by performing <u>tricks</u> or just by being <u>watched</u> by the visitors. Animals are also <u>hunted</u> for fun — deer, foxes and game birds are all hunted, and sometimes <u>dogs</u> are used to track them down and kill them.

Companionship

People enjoy keeping lots of different types of animals as <u>pets</u>. <u>Dogs</u> and <u>cats</u> are probably the most common pets — they can give people <u>company</u>, and can be <u>loyal</u> and <u>affectionate</u>.

Medicinal uses

Animals can be used to produce <u>antibodies</u> for use in vaccines, and genetic engineering has enabled us to breed animals that will produce certain <u>drugs</u> for human use (normally in their <u>milk</u>). Scientists are researching the possibility of <u>transplanting organs</u> from animals such as pigs into humans, if an appropriate human organ isn't available. Animals are also widely used for <u>testing drugs</u> before they are released for human use.

Aaaaw.
Puppy wuppy.

Do Animals Have Rights?

People who believe that animals should have the same <u>rights</u> as humans might <u>protest</u> about the following:

1) The use of animals in <u>medical</u> or <u>cosmetic research</u> that puts them in any pain or discomfort.
2) <u>Hunting</u>, particularly for <u>sport</u>, but also for <u>food</u> where it's considered cruel or unnecessary.
3) Using animals for <u>entertainment</u> in circuses etc. Some people are also against <u>zoos</u>.
4) Using animals to provide <u>unnecessary luxuries</u>, e.g. fur coats and ivory ornaments.
5) <u>Intensive farming</u>, where animals are mistreated or kept in cramped conditions.

Of course, different people might say there are other things to consider too...

1) If drugs weren't <u>tested</u> on animals, potentially <u>unsafe</u> drugs could get onto the market.
2) Some say that hunting is a humane way to keep foxes <u>under control</u> so they don't kill farm animals.
3) Circus owners claim that the animals are treated <u>kindly</u> and actually <u>enjoy performing</u>.
4) Zoos <u>educate</u> people and many have <u>breeding programmes</u> aimed at helping <u>endangered species</u>.
5) Where do we <u>draw the line</u>? Should we not kill things that carry <u>disease</u>? Is using <u>fly spray</u> cruel?

Stick up for animal rights, or there won't be animal lefts...

We often attribute human feelings to animals (this is called <u>anthropomorphism</u>), but is it right to? Just because we might feel afraid or stressed in a situation doesn't mean an animal would. On the other hand, we shouldn't assume it wouldn't, either — humans and animals do have behaviours <u>in common</u>.

Revision Summary for Section Twelve

What an interesting topic. I don't know about you, but I'm fascinated by this sort of stuff. There are so many things for you to try out at home — conditioning your pets to salivate when you ring a bell, checking if your brothers and sisters are self-aware, habituating your parents to loud music... I could go on. There are also a lot of issues in this topic, so make sure you know both sides of the argument when it comes to animal testing, hunting and suchlike. It doesn't matter what your own opinion is — you've got to show that you're aware of other people's viewpoints too.

Anyway, I digress. Back to the important issue of finding out how much of this stuff you've taken in.

1) Define 'behaviour'. Why do we need behaviour? What influences it?
2) Name three reflex actions. What are reflexes for?
3) Which animal shows negative phototaxis? What is this?
4) What's the process called when birds learn to ignore scarecrows? Give another example.
5) Which type of conditioning did Ivan Pavlov study? What experiment did Pavlov do?
6) Give a definition of classical conditioning. Make sure you use the word 'passively'.
7) Who trained rats and pigeons to press levers for rewards in a special type of box?
8) Name three species which communicate through sound.
9) What are pheromones? What uses them and why would they want to?
10) What is the 'waggle dance'? Name two other behaviour signals used by animals.
11) Which species of animal uses really complex language to communicate?
12) Name two types of non-verbal communication that can be automatic and unconscious.
13) Give two different definitions of self-awareness. Which one can you test using a mirror?
14) Why do herbivores need to spend so long feeding every day?
15) Name two ways in which herbivores have adapted so that they avoid being caught by predators.
16) How are carnivores generally adapted for a life of hunting?
17) Which tend to hunt individually — animals that eat large prey or animals that eat small prey?
18) Why do herring gulls have a red spot on their beaks?
19) Name two species that use tools (apart from humans, of course). What tools do they use?
20) Why do females tend to be choosy when it comes to selecting a mate?
21) Why are courtship displays species-specific?
22) Describe some of the mating patterns found in the animal kingdom. How common is monogamy?
23) What's so special about how weaver birds protect their young?
24) Apart from feeding and protecting, what do some animals do as part of caring for their young?
25) Briefly explain the evolutionary argument for why mammals care for their young.
26) Describe four ways in which earthworms improve soil fertility.
27) Name two advantages and two disadvantages of living in water.
28) What water-related problem faces salmon? How do they overcome it?
29) How do amoebas regulate their water content?
30) Explain what an algal bloom is.
31) Why do algal blooms normally occur in late spring to late summer?
32) What role does an adult amphibian's skin play in gas exchange?
33) What do we think is the oldest known human ancestor? When does this organism date from?
34) Name five animals that humans have exploited during their development into modern man.
35) How are animals used by humans today (apart from to provide meat and farm labour)?
36) Give arguments in favour of zoos, hunting and animal-testing.
37) What is anthropomorphism? How is this applicable to the animal-rights debate?

Answering Experiment Questions (i)

You'll definitely get some questions in the exam about experiments. They can be about any topic under the Sun — but if you learn the basics and throw in a bit of common sense, you'll be fine.

Read the Question *Carefully*

The question might describe an <u>experiment</u>, e.g. —

> Ellie had three different powdered fertilisers: A, B and C.
> She investigated which fertiliser was most effective when growing pea plants.
> Ellie planted a seed in each of 12 pots. She then added fertiliser A to three pots, fertiliser B to three pots and fertiliser C to three. She did not add any fertiliser to the remaining three pots.
>
>
> Fertiliser A   Fertiliser B   Fertiliser C   No fertiliser
>
> She watered each pot daily and after three weeks she measured the heights of each seedling.

1. What is the independent variable?
   The type of fertiliser.

   The <u>independent variable</u> is the thing that is <u>changed</u>.

   The <u>dependent variable</u> is the thing that's <u>measured</u>.

2. What is the dependent variable?
   The heights of the seedlings.

   To make it a <u>fair test</u>, you've got to keep <u>all</u> the other variables the same. Or else you won't know if the <u>only thing</u> affecting the dependent variable is the <u>independent variable</u>.

   There are <u>loads</u> of other things that must be kept the same for each pot in this experiment. You could also have put <u>temperature</u> or the <u>type of soil</u>, etc.

3. Give two variables that must be kept the same to make it a fair test.

   1. The amount of fertiliser.

   2. The amount of light.

   It's easy to keep the variables the same in this experiment as it's in a <u>laboratory</u>. But it can sometimes be <u>trickier</u>. For example, if the seeds were growing in <u>fields</u>, it'd be hard to make sure that they all had exactly the same soil conditions, and got the same amount of water and light, etc.

   It's even harder to make investigations involving <u>people</u> fair.

   If, say, the effect of a person's age on their blood pressure was being investigated, there'd be <u>loads of other variables</u> to consider — weight, diet and whether someone's a smoker could make a <u>big difference</u> to their blood pressure.

4. What is the control group in this investigation?

   The group of pots with no fertiliser.

   To make it a <u>fairer</u> test, it would be better if just nonsmokers with a similar weight and diet were used.

A <u>control group</u> isn't really part of the experiment, but it's kept in the <u>same conditions</u> as the rest of the experiment. You can compare changes in the experiment with those that happened to the control group, and see if the changes might have happened <u>anyway</u>. Control groups make results <u>more meaningful</u>.

In this experiment the seeds might <u>grow better</u> without any fertiliser — with a control group you can check for this.

Control groups are used when <u>testing drugs</u>. People can feel better just because they've been given a drug that they <u>believe</u> will work. To rule this out, researchers give one group of patients <u>dummy pills</u> (called placebos) — but they <u>don't tell them</u> that their pills aren't the real thing. This is the control group. By doing this, they can tell if the real drug is actually working.

Answering Experiment Questions (ii)

5. Why was each type of fertiliser added to three pots, instead of just one?

To check for anomalous results and make

the results more <u>reliable</u>.

6. The table below shows the heights of the seedlings in each pot.

	First pot	Second pot	Third pot	Mean
Fertiliser A	4.4 cm	5.2 cm	4.2 cm	
Fertiliser B	8.3 cm	7.9 cm	8.7 cm	8.3 cm
Fertiliser C	6.7 cm	5.7 cm	(0 cm)	6.2 cm
No fertiliser	2.4 cm	1.9 cm	2.6 cm	2.3 cm

When an experiment is <u>repeated</u>, the results will usually be <u>slightly different</u> each time.

The <u>mean</u> (or average) of the measurements is usually used to represent the values.

The more times the experiment is <u>repeated</u> the <u>more reliable</u> the average will be.

To find the mean:

> ADD TOGETHER all the data values and DIVIDE by the total number of values in the sample.

The <u>range</u> is how far the data <u>spreads</u>.

You just work out the <u>difference</u> between the <u>highest</u> and <u>lowest</u> numbers.

a) Calculate the mean height of the seedlings grown with fertiliser A.

$Mean = (4.4 + 5.2 + 4.2) \div 3 = \underline{4.6\ cm}$

b) What is the range of the heights of the seedlings grown with fertiliser A?

$5.2 - 4.2 = \underline{1.0\ cm}$

If one of the results doesn't seem to fit in, it's called an <u>anomalous</u> result. You should usually <u>ignore</u> an anomalous result. It's been <u>ignored</u> when the mean was worked out.

This is a <u>random error</u> — it only happens occasionally.

7. One of the results in the table is anomalous. Circle the result and suggest why it may have occurred.

The seed may have had something

genetically wrong with it.

If the same mistake is made every time, it's a <u>systematic error</u>, e.g. if you measured from the very end of your ruler instead of from the 0 cm mark every time, meaning <u>all</u> your measurements would be a <u>bit small</u>.

8. What conclusion can you draw from these results?

Fertiliser B makes pea plants grow

taller over the first three weeks than

fertilisers A or C, given daily watering.

Be careful that your conclusions <u>match</u> the data you've got, and <u>don't</u> go any further.

You can't say that fertiliser B will always be better than fertilisers A or C, because:

- The results may be <u>totally different</u> with <u>another type of plant</u>.
- After four weeks, the plants grown with fertiliser B may all <u>drop dead</u>, while the others <u>keep growing</u>. Etc.

Mistakes happen...

NASA made a bit of a boo-boo once. They muddled up measurements in pounds and newtons and caused the Mars Climate Orbiter to burn up in the Martian atmosphere. Ooops. It just goes to show that anyone can make a mistake, even a bunch of brainy boffins. So always double-check everything.

Answering Experiment Questions (iii)

Use Sensible Measurements for Your Variables

Pu-lin did an experiment to see how the mass of a potato changed depending on the sugar solution it was in.

She started off by making potato tubes 5 cm in length, 1 cm in diameter and 2.0 g in mass. She then filled a beaker with 500 ml of pure water and placed a potato tube in it for 30 minutes. She repeated the experiment with different amounts of sugar dissolved in the water. For each potato tube, she measured the new mass. She did the experiment using Charlotte, Desiree, King Edward and Maris Piper potatoes.

Before she started, she did a trial run, which showed that most of the potato tubes shrunk to a minimum of 1 g (in a really strong sugar solution) or grew to a maximum of 3 g (in pure water).

1. What kind of variable was the list of potatoes?

   A A continuous variable ☐

   B A categoric variable ☑

   C An ordered variable ☐

   D A discrete variable ☐

2. Pu-lin should add sugar in intervals of...

   A a pinch ☐

   B a teaspoon ☑

   C a cupful ☐

   D a bucketful ☐

3. The balance used to find the mass of the potato should be capable of measuring...

   A to the nearest 0.01 gram ☑

   B to the nearest 0.1 gram ☐

   C to the nearest gram ☐

   D to the nearest 10 grams ☐

Categoric variables are variables that can't be related to size or quantity — they're types of things. E.g. names of potatoes or types of fertiliser.

Continuous data is numerical data that can have any value within a range — e.g. length, volume, temperature and time.

Note: You can't measure the exact value of continuous data. Say you measure a height as 5.6 cm to the nearest mm. It's not exact — you get a more precise value if you measure to the nearest 0.1 mm or 0.01 mm, etc.

Ordered variables are things like small, medium and large lumps, or warm, very warm and hot.

Discrete data is the type that can be counted in chunks, where there's no in-between value. E.g. number of people is discrete, not continuous, because you can't have half a person.

It's important to use sensible values for variables.

It's no good using loads of sugar or really weedy amounts like a pinch at a time cos you'd be there forever and the results wouldn't show any significant difference. (You'd get different amounts of sugar in each pinch anyway.)

A balance measuring only to the nearest gram, or bigger, would not be sensitive enough — the changes in mass are likely to be quite small, so you'd need to measure to the nearest 0.01 gram to get the most precise results.

The sensitivity of an instrument is the smallest change it can detect, e.g. some balances measure to the nearest gram, but really sensitive ones measure to the nearest hundredth of a gram.

For measuring tiny changes — like from 2.00 g to 1.92 g — the more sensitive balance is needed.

You also have to think about the precision and accuracy of your results.

Precise results are ones taken with sensitive instruments, e.g. volume measured with a burette will be more precise than volume measured with a 100 ml beaker. Really accurate results are those that are really close to the true answer. It's possible for results to be precise but not very accurate, e.g. a fancy piece of lab equipment might give results that are precise, but if it's not calibrated properly those results won't be accurate.

I take my tea milky with two bucketfuls of sugar... mmm...

Accuracy, precision and sensitivity are difficult things to get your head around — a sensitive piece of equipment is likely to give precise results (but not necessarily very accurate results). If the equipment is used properly and calibrated well then the results are more likely to be accurate...

Answering Experiment Questions (iv)

Once you've collected all your data together, you need to analyse it to find any relationships between the variables. The easiest way to do this is to draw a graph, then describe what you see...

Graphs Are Used to Show Relationships

These are the results Pu-lin obtained with the King Edward potato.

Number of teaspoons of sugar	0	2	4	6	8	10	12	14	16	18	20
Mass of potato tube (g)	2.50	2.40	2.23	2.10	2.02	1.76	1.66	1.25	1.47	1.3	1.15

4. a) Nine of the points are plotted below.
    Plot the remaining **two** points on the graph.

To plot the points, use a <u>sharp</u> pencil and make a <u>neat</u> little cross.

nice clear mark

smudged unclear marks

   b) Draw a straight line of best fit for the points.

A line of best fit is drawn so that it's easy to see the <u>relationship</u> between the variables. You can then use it to <u>estimate</u> other values.

When drawing a line of best fit, try to draw the line through or as near to as many points as possible, ignoring any <u>anomalous</u> results.

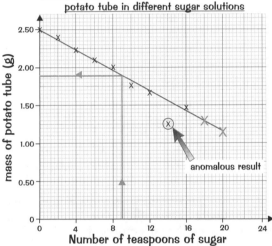

Scattergram to show the mass of a King Edward potato tube in different sugar solutions

anomalous result

This is a <u>scattergram</u> — they're used to see if two variables are <u>related</u>.

5. Estimate the mass of the potato tube if you added nine teaspoons of sugar.

    Estimate of mass = ___1.90 g (see graph)___

This graph shows a <u>negative correlation</u> between the variables. This is where one variable <u>increases</u> as the other one <u>decreases</u>.

The other correlations you could get are:

<u>Positive correlation</u> — this is where as one variable <u>increases</u> so does the other one.

<u>No correlation</u> — this is where there's <u>no obvious relationship</u> between the variables.

6. What can you conclude from these results?

    There is a negative correlation between the number of teaspoons of sugar and the mass of potato tube. Each additional teaspoon causes the potato tube to lose mass.

In lab-based experiments like this one, you can say that one variable <u>causes</u> the other one to change. The extra sugar <u>causes</u> the potato to lose mass. You can say this because everything else has <u>stayed the same</u> — nothing else could be causing the change.

There's a positive correlation between revising and good marks...

...really, it's true. Other ways to improve your marks are to practise plotting graphs, and learning how to read them properly — make sure you're reading off the right axis for a start, and don't worry about drawing lines on the graph if it helps you to read it. Always double-check your answer... just in case.

Answering Experiment Questions (v)

Not all experiments can be carefully controlled in a laboratory. Some have to be done in the real world. Unfortunately, this creates complications of its own.

Relationships Do NOT Always Tell Us the Cause

Melanomas are a dangerous form of skin cancer. It's thought that UV damage may increase the risk of getting skin cancer later in life, so people are advised to avoid being in direct sunlight for long periods at a time.

The graph shows the number of new cases of melanoma found per year in people who spend at least 5 hours of each working day exposed to direct sunlight.

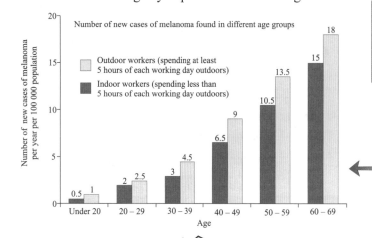

In large studies done outside a lab it's really difficult to keep all the variables the same and to make sure the control group are kept in the same conditions.

In this study the control group are people who work indoors.

This is a bar chart. It contains a key to tell you which colour bars relate to which group.

1. The graph is representative of a country that has 500 000 people aged between 40 and 49 who work outdoors.

   Using the graph, estimate how many of these people will get melanomas in a given year.

   9 × 5 = 45 people

Here they're asking you to estimate the number of people out of 500 000 — the graph tells you the number of people in 100 000. Don't get caught out — read the question really carefully.

2. What conclusion can you draw from these results?

   Melanomas are found more frequently in people who spend at least 5 hours of each working day in direct sun than in people who work indoors.

When describing the data and drawing conclusions it's really important that you don't say that working out in the sun causes melanomas. The graph only shows that there's a positive correlation between the two.

In studies like this where you're unable to control everything, it's possible a third variable is causing the relationship. For example, many people who work outside are farm labourers or builders. Pesticides and other toxic chemicals that you might be exposed to in these professions could cause the increased rate of melanomas.

3. Suggest how the data may have been collected.

   e.g. from medical records

Use your common sense to think of a sensible answer.

Try to suggest a method to get reliable results. For example, it's very unlikely that the data would have been collected by a telephone survey or an internet search.

Who wants to live in the real world anyway... (I have my very own special one)

It's really difficult to prove what causes what in science, especially with all the things you've got to control. The experiments are usually done in a lab first so that you can control as much as possible. Then they're done in the real world to see if the same thing happens, and to find any unexpected results.

Index

Index

Index

Index and Answers

Answers

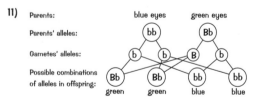